INVESTIGATING EARTH SYSTEMS™

AN INQUIRY EARTH SCIENCE PROGRAM

INVESTIGATING CLIMATE AND WEATHER

Michael J. Smith Ph.D.
American Geological Institute

John B. Southard Ph.D.
Massachusetts Institute of Technology

Colin Mably
Curriculum Developer

Developed by the American Geological Institute
Supported by the National Science Foundation and
the American Geological Institute Foundation

Published by
It's About Time Inc., Armonk, NY

It's About Time, Inc.
84 Business Park Drive, Armonk, NY 10504
Phone (914) 273-2233 Fax (914) 273-2227
Toll Free (888) 698-TIME
www.Its-About-Time.com

President
Laurie Kreindler

Project Editor	**Creative Artwork**	**Senior Photo Consultant**
Ruta Demery	Dennis Falcon	Bruce F. Molnia
Design	**Safety Reviewer**	**Photo Research**
John Nordland	Dr. Ed Robeck	Caitlin Callahan
		Eric Shih
Studio Manager	**Production**	
Jon Voss	Burmar Technical Corporation	**Contributing Writer**
Associate Editor	**Technical Art**	William Jones
Al Mari	Armstrong/Burmar	

Illustrations and Photos

C23, American Meteorological Society; C10, Annapolis Weathervanes; C54, illustration by Stuart Armstrong; C6, C7, C23, C25, C26, C27, C33, C34, C35, C39, C50, C51, C52, C61, C70, C75, C83, C84, C85, illustrations by Burmar Technical Corporation; C12, Cody Mercantile Catalog; C65 (left), Corbis Royalty Free Images; C24, C30, C31, C40, C42, C43, The DataStreme Project, American Meteorological Society; C11 (2nd down, right), (2nd from bottom right), Digital Royalty Free Images; C44, Digital Vision Royalty Free Images; C66 (bottom), Digital Vision Royalty Free Images; Cv, Cxii, C2, C15, C25, C28, C41, C49, C63, C70, C82, illustrations by Dennis Falcon; C72, (bottom), Geoff Hargreaves, USGS/National Ice Core Laboratory; C67, John Karachewski; C11 (2nd from bottom, left), (top right), James Koermer, Plymouth State College; C11, (top left) Ralph Kresge, NOAA; C73 (bottom), Laboratory of Tree-Ring Research, University of Tennessee; C71, Steven Manchester; courtesy of Oregon Department of Geology and Mineral Industries; C65, Martin Miller; C11 (bottom photos), C56, C66, (top), C69, C72 (top), C79, C85, Bruce F. Molnia; C11 (2nd from top, left), C28, C45, Joe Moran; C37, NASA; C22, NOAA; C1, C8, C14, C20, C48, C57 (top, bottom), C73 (top), C76, C77, PhotoDisc; C9, C18, Doug Sherman, Geo File Photography; C54, source, USGS

All student activities in this textbook have been designed to be as safe as possible, and have been reviewed by professionals specifically for that purpose. As well, appropriate warnings concerning potential safety hazards are included where applicable to particular activities. However, responsibility for safety remains with the student, the classroom teacher, the school principals, and the school board.

Investigating Earth Systems™ is a registered trademark of the American Geological Institute. Registered names and trademarks, etc. used in this publication, even without specific indication thereof, are not to be considered unprotected by law.

It's About Time® is a registered trademark of It's About Time, Inc. Registered names and trademarks, etc., used in this publication, even without specific indication thereof, are not to be considered unprotected by law.

© Copyright 2001: American Geological Institute

All rights reserved. No part of this publication may be reproduced, stored in a retrieval system, or transmitted, in any form or by any means, electronic, mechanical, photocopying, recording, or otherwise, without the prior written permission of the copyright owner.

Care has been taken to trace the ownership of copyright material contained in this publication. The publisher will gladly receive any information that will rectify any reference or credit line in subsequent editions.

Printed and bound in the United States of America

Soft Cover ISBN #1-58591-078-3   Hard Cover ISBN #1-58591-112-7

2 3 4 5 QC 05 04 03

This project was supported, in part, by the
National Science Foundation (grant no. 9353035)

Opinions expressed are those of the authors and not necessarily those of the National Science Foundation or the donors of the American Geological Institute Foundation.

Acknowledgements

Principal Investigator

Michael Smith is Director of Education at the American Geological Institute in Alexandria, Virginia. Dr. Smith worked as an exploration geologist and hydrogeologist. He began his Earth Science teaching career with Shady Side Academy in Pittsburgh, PA in 1988, and most recently taught Earth Science at the Charter School of Wilmington, DE. He earned a doctorate from the University of Pittsburgh's Cognitive Studies in Education Program and joined the faculty of the University of Delaware School of Education in 1995. Dr. Smith received the Outstanding Earth Science Teacher Award for Pennsylvania from the National Association of Geoscience Teachers in 1991, served as Secretary of the National Earth Science Teachers Association, and is a reviewer for Science Education and The Journal of Research in Science Teaching. He worked on the Delaware Teacher Standards, Delaware Science Assessment, National Board of Teacher Certification, and AAAS Project 2061 Curriculum Evaluation programs.

Senior Writer

John Southard received his undergraduate degree from the Massachusetts Institute of Technology in 1960 and his doctorate in geology from Harvard University in 1966. After a National Science Foundation postdoctoral fellowship at the California Institute of Technology, he joined the faculty at the Massachusetts Institute of Technology, where he is currently Professor of Geology emeritus. He was awarded the MIT School of Science teaching prize in 1989 and was one of the first cohorts of first MacVicar Fellows at MIT, in recognition of excellence in undergraduate teaching. He has taught numerous undergraduate courses in introductory geology, sedimentary geology, field geology, and environmental Earth science, both at MIT and in Harvard's adult education program. He was editor of the Journal of Sedimentary Petrology from 1992 to 1996, and he continues to do technical editing of scientific books and papers for SEPM, a professional society for sedimentary geology. Dr. Southard received the 2001 Neil Miner Award from the National Association of Geoscience Teachers.

Project Director/Curriculum Designer

Colin Mably has been a key curriculum developer for several NSF-supported national curriculum projects. As learning materials designer to the American Geological Institute, he has directed the design and development of the IES curriculum modules and also training workshops for pilot and field-test teachers.

INVESTIGATING CLIMATE AND WEATHER

Project Team

Marcus Milling
Executive Director - AGI, VA

Michael Smith
Principal Investigator - Director of Education - AGI, VA

Colin Mably
Project Director/Curriculum Designer - Educational Visions, MD

Fred Finley
Project Evaluator
University of Minnesota, MN

Lynn Lindow
Pilot Test Evaluator
University of Minnesota, MN

Harvey Rosenbaum
Field Test Evaluator
Montgomery School District, MD

Ann Benbow
Project Advisor - American Chemical Society, DC

Robert Ridky
Original Project Director
University of Maryland, MD

Chip Groat
Original Principal Investigator
University of Texas
El Paso, TX

Marilyn Suiter
Original Co-principal Investigator - AGI, VA

William Houston
Project Manager

Eric Shih - Project Assistant

Original and Contributing Authors

Oceans
George Dawson
Florida State University, FL

Joseph F. Donoghue
Florida State University, FL

Ann Benbow
American Chemical Society

Michael Smith
American Geological Institute

Soil
Robert Ridky
University of Maryland, MD

Colin Mably - LaPlata, MD

John Southard
Massachusetts Institute of Technology, MA

Michael Smith
American Geological Institute

Fossils
Robert Gastaldo
Colby College, ME

Colin Mably - LaPlata, MD

Michael Smith
American Geological Institute

Climate and Weather
Mike Mogil
How the Weather Works, MD

Ann Benbow
American Chemical Society

Michael Smith
American Geological Institute

Energy Resources
Laurie Martin-Vermilyea
American Geological Institute

Michael Smith
American Geological Institute

Dynamic Planet
Michael Smith
American Geological Institute

Rocks and Landforms
Michael Smith
American Geological Institute

Water as a Resource
Ann Benbow
American Chemical Society

Michael Smith
American Geological Institute

Materials and Minerals
Mary Poulton
University of Arizona, AZ

Colin Mably - LaPlata, MD

Michael Smith
American Geological Institute

Advisory Board

Jane Crowder
Middle School Teacher, WA

Kerry Davidson
Louisiana Board of Regents, LA

Joseph D. Exline
Educational Consultant, VA

Louis A. Fernandez
California State University, CA

Frank Watt Ireton
National Earth Science Teachers Association, DC

LeRoy Lee
Wisconsin Academy of Sciences, Arts and Letters, WI

Donald W. Lewis
Chevron Corporation, CA

James V. O'Connor (deceased)
University of the District of Columbia, DC

Roger A. Pielke Sr.
Colorado State University, CO

Dorothy Stout
Cypress College, CA

Lois Veath
Advisory Board Chairperson
Chadron State College, NE

Pilot Test Teachers

Debbie Bambino
Philadelphia, PA

Barbara Barden - Rittman, OH

Louisa Bliss - Bethlehem, NH

Mike Bradshaw - Houston TX

Greta Branch - Reno, NV

Garnetta Chain - Piscataway, NJ

Roy Chambers Portland, OR

Laurie Corbett - Sayre, PA

James Cole - New York, NY

Collette Craig - Reno, NV

Anne Douglas - Houston, TX

Jacqueline Dubin - Roslyn, PA

Jane Evans - Media, PA

Gail Gant - Houston, TX

Joan Gentry - Houston, TX

Pat Gram - Aurora, OH

Robert Haffner - Akron, OH

Joe Hampel - Swarthmore, PA

Wayne Hayes - West Green, GA

Mark Johnson - Reno, NV

Cheryl Joloza - Philadelphia, PA

Jeff Luckey - Houston, TX

Karen Luniewski
Reistertown, MD

Cassie Major - Plainfield, VT

Carol Miller - Houston, TX

Melissa Murray - Reno, NV

Mary-Lou Northrop
North Kingstown, RI

Keith Olive - Ellensburg, WA

Tracey Oliver - Philadelphia, PA

Nicole Pfister - Londonderry, VT

Beth Price - Reno, NV

Joyce Ramig - Houston, TX

Julie Revilla - Woodbridge, VA

Steve Roberts - Meredith, NH

Cheryl Skipworth
Philadelphia, PA

Brent Stenson - Valdosta, GA

Elva Stout - Evans, GA

Regina Toscani
Philadelphia, PA

Bill Waterhouse
North Woodstock, NH

Leonard White
Philadelphia, PA

Paul Williams - Lowerford, VT

Bob Zafran - San Jose, CA

Missi Zender - Twinsburg, OH

Field Test Teachers

Eric Anderson - Carson City, NV

Katie Bauer - Rockport, ME

Kathleen Berdel - Philadelphia, PA

Wanda Blake - Macon, GA

Beverly Bowers
Mannington, WV

Rick Chiera - Monroe Falls, OH

Don Cole - Akron, OH

Patte Cotner - Bossier City, LA

Johnny DeFreese - Haughton, LA

Mary Devine - Astoria, NY

Cheryl Dodes - Queens, NY

Brenda Engstrom - Warwick, RI

Lisa Gioe-Cordi - Brooklyn, NY

Pat Gram - Aurora, OH

Mark Johnson - Reno, NV

Chicory Koren - Kent, OH

Marilyn Krupnick
Philadelphia, PA

Melissa Loftin - Bossier City, LA

Janet Lundy - Reno, NV

Vaughn Martin - Easton, ME

Anita Mathis - Fort Valley, GA

Laurie Newton - Truckee, NV

Debbie O'Gorman - Reno, NV

Joe Parlier - Barnesville, GA

Sunny Posey - Bossier City, LA

Beth Price - Reno, NV

Stan Robinson
Mannington, WV

Mandy Thorne
Mannington, WV

Marti Tomko
Westminster, MD

Jim Trogden - Rittman, OH

Torri Weed - Stonington, ME

Gene Winegart - Shreveport, LA

Dawn Wise - Peru, ME

Paula Wright - Gray, GA

IMPORTANT NOTICE

The *Investigating Earth Systems*™ series of modules is intended for use by students under the direct supervision of a qualified teacher. The experiments described in this book involve substances that may be harmful if they are misused or if the procedures described are not followed. Read cautions carefully and follow all directions. Do not use or combine any substances or materials not specifically called for in carrying out experiments. Other substances are mentioned for educational purposes only and should not be used by students unless the instructions specifically indicate.

The materials, safety information, and procedures contained in this book are believed to be reliable. This information and these procedures should serve only as a starting point for classroom or laboratory practices, and they do not purport to specify minimal legal standards or to represent the policy of the American Geological Institute. No warranty, guarantee, or representation is made by the American Geological Institute as to the accuracy or specificity of the information contained herein, and the American Geological Institute assumes no responsibility in connection therewith. The added safety information is intended to provide basic guidelines for safe practices. It cannot be assumed that all necessary warnings and precautionary measures are contained in the printed material and that other additional information and measures may not be required.

This work is based upon work supported by the National Science Foundation under Grant No. 9353035 with additional support from the Chevron Corporation. Any opinions, findings, and conclusions or recommendations expressed in this publication are those of the authors and do not necessarily reflect the views of the National Science Foundation or the Chevron Corporation. Any mention of trade names does not imply endorsement from the National Science Foundation or the Chevron Corporation.

Table of Contents

Introducing Climate and Weather	Cxi
Why Are Climate and Weather Important?	Cxii
Investigation 1: Observing Weather	C1
Elements of Weather	C8
Investigation 2: Comparing Weather Reports	C14
Weather Reports and Forecasts	C18
Investigation 3: Weather Maps	C22
Weather Maps	C30
Investigation 4: Weather Radiosondes, Satellites, and Radar	C37
The Weather High in the Atmosphere	C44
Investigation 5: The Causes of Weather	C48
The Water Cycle	C54
Investigation 6: Climates	C60
Weather and Climate	C65
Investigation 7: Exploring Climate Change	C69
Measuring Climate Change	C74
Investigation 8: Climate Change Today	C79
Global Climate Change	C83
Reflecting	C87
The Big Picture	C88
Glossary	C89

INVESTIGATING CLIMATE AND WEATHER

Using Investigating Earth Systems

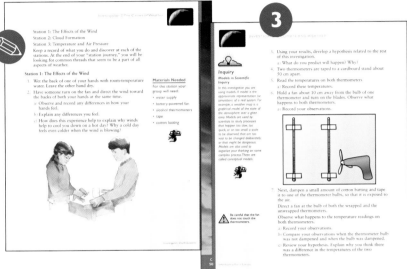

Look for the following features in this module to help you learn about the Earth system.

1. Key Question
Before you begin, you will be asked to think about the key question you will investigate. You do not need to come up with a correct answer. Instead you will be expected to take some time to think about what you already know. You can then share your ideas with your small group and with the class.

2. Investigate
Geoscientists learn about the Earth system by doing investigations. That is exactly what you will be doing. Sometimes you will be given the procedures to follow. Other times you will need to decide what question you want to investigate and what procedure to follow.

3. Inquiry
You will use inquiry processes to investigate and solve problems in an orderly way. Look for these reminders about the processes you are using.

Throughout your investigations you will keep your own journal. Your journal is like one that scientists keep when they investigate a scientific question. You can enter anything you think is important during the investigation. There will also be questions after many of the **Investigate** steps for you to answer and enter in your journal. You will also need to think about how the Earth works as a set of systems. You can write the connections you make after each investigation on your *Earth System Connection* sheet in your journal.

4. Digging Deeper
Scientists build on knowledge that others have discovered through investigation. In this section you can read about the insights scientists have about the question you are investigating. The questions in **As You Read** will help you focus on the information you are looking for.

5. Review and Reflect
After you have completed each investigation, you will be asked to reflect on what you have learned and how it relates to the "Big Picture" of the Earth system. You will also be asked to think about what scientific inquiry processes you used.

6. Investigation: Putting It All Together
In the last investigation of the module you will have a chance to "put it all together." You will be asked to apply all that you have learned in the previous investigations to solve a practical problem. This module is just the beginning! You continue to learn about the Earth system every time you ask questions and make observations about the world around you.

INVESTIGATING CLIMATE AND WEATHER

The Earth System

The Earth System is a set of systems that work together in making the world we know. Four of these important systems are:

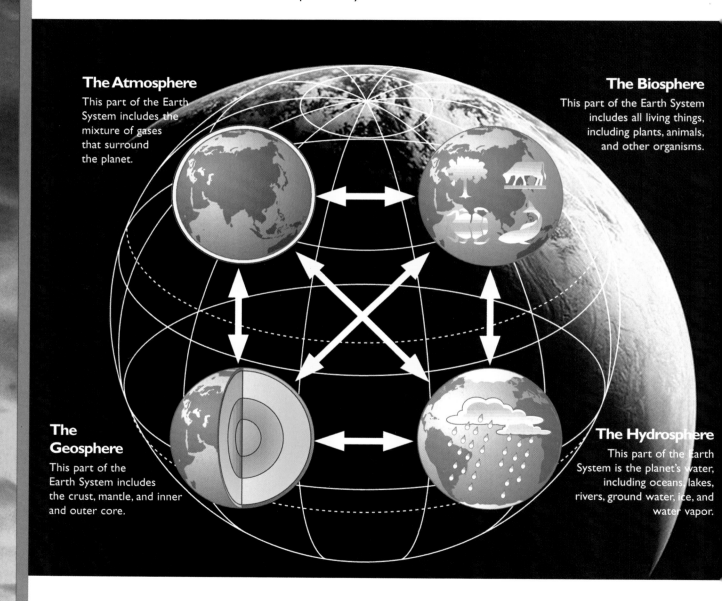

The Atmosphere
This part of the Earth System includes the mixture of gases that surround the planet.

The Biosphere
This part of the Earth System includes all living things, including plants, animals, and other organisms.

The Geosphere
This part of the Earth System includes the crust, mantle, and inner and outer core.

The Hydrosphere
This part of the Earth System is the planet's water, including oceans, lakes, rivers, ground water, ice, and water vapor.

These systems, and others, have been working together since the Earth's beginning about 4.5 billion years ago. They are still working, because the Earth is always changing, even though we cannot always observe these changes. Energy from within and outside the Earth leads to changes in the Earth System. Changes in any one of these systems affects the others. This is why we think of the Earth as made of interrelated systems.

During your investigations, keep the Earth System in mind. At the end of each investigation you will be asked to think about how the things you have discovered fit with the Earth System.

To further understand the Earth System, take a look at THE BIG PICTURE shown on page 88.

INVESTIGATING CLIMATE AND WEATHER

Introducing Inquiry Processes

When geologists and other scientists investigate the world, they use a set of inquiry processes. Using these processes is very important. They ensure that the research is valid and reliable. In your investigations, you will use these same processes. In this way, you will become a scientist, doing what scientists do. Understanding inquiry processes will help you to investigate questions and solve problems in an orderly way. You will also use inquiry processes in high school, in college, and in your work.

During this module, you will learn when, and how, to use these inquiry processes. Use the chart below as a reference about the inquiry processes.

Inquiry Processes:	How scientists use these processes
Explore questions to answer by inquiry	Scientists usually form a question to investigate after first looking at what is known about a scientific idea. Sometimes they predict the most likely answer to a question. They base this prediction on what they already know to be true.
Design an investigation	To make sure that the way they test ideas is fair, scientists think very carefully about the design of their investigations. They do this to make sure that the results will be valid and reliable.
Conduct an investigation	After scientists have designed an investigation, they conduct their tests. They observe what happens and record the results. Often, they repeat a test several times to ensure reliable results.
Collect and review data using tools	Scientists collect information (data) from their tests. The data may be numerical (numbers), or verbal (words). To collect and manage data, scientists use tools such as computers, calculators, tables, charts, and graphs.
Use evidence to develop ideas	Evidence is very important for scientists. Just as in a court case, it is proven evidence that counts. Scientists look at the evidence other scientists have collected, as well as the evidence they have collected themselves.
Consider evidence for explanations	Finding strong evidence does not always provide the complete answer to a scientific question. Scientists look for likely explanations by studying patterns and relationships within the evidence.
Seek alternative explanations	Sometimes, the evidence available is not clear or can be interpreted in other ways. If this is so, scientists look for different ways of explaining the evidence. This may lead to a new idea or question to investigate.
Show evidence & reasons to others	Scientists communicate their findings to other scientists to see if they agree. Other scientists may then try to repeat the investigation to validate the results.
Use mathematics for science inquiry	Scientists use mathematics in their investigations. Accurate measurement, with suitable units is very important for both collecting and analyzing data. Data often consist of numbers and calculations.

Introducing Climate and Weather

Have you ever been in the middle of a powerful storm?

Have you ever wondered where the information for weather reports comes from and why weather forecasts are important?

Have you ever seen the effects of a serious lack of rain?

Have you ever seen clouds forming over a body of water?

INVESTIGATING CLIMATE AND WEATHER

Why Are Climate and Weather Important?

"What's the weather going to be like today?" That is often the first question on your mind when you wake up in the morning. What you wear, and sometimes even what you do on any given day depends on the weather. Weather can change very quickly. It could be sunny and warm in the morning, and in the afternoon you could be faced with dangerous thunderstorms and even tornadoes. You count on meteorologists (scientists who study the weather) to provide you with daily weather information.

On the other hand, you depend on the climate to give you fairly similar weather conditions year after year. Farmers expect the same length of growing season each year. Ski-resort operators anticipate a reasonable snowfall each year. They rely on the climate in the area to remain the same. Yet over very long periods of time, climate can change. Climatologists (scientist who study the climate) have evidence that the climate has changed many times in the past.

What Will You Investigate?

These investigations will put you in the roles of weather reporter, fact finder, and inquiring student. You will be using weather instruments and observations to make weather maps, weather reports, and weather forecasts. You will look for patterns in your weather data and explore reasons for those patterns. You will explore how climate has changed over time, the effects of climate on your life now, and what might happen if the climate changes in the future.

Here are some of the things that you will investigate:

- how weather instruments work;
- what is contained in a weather report and map;
- how weather observations are made;
- the underlying causes of weather patterns;
- the difference between climate and weather;
- how scientists know that the climate has changed in the past;
- how climate is changing now.

Investigation 1: Observing Weather

Investigation 1:
Observing Weather

 Key Question
Before you begin, first think about this key question.

How is weather observed?

Think about what you know about weather reports and weather maps. What sort of information goes into one? How is this information obtained? Make a list that combines what you know about obtaining weather information with questions that you might be able to answer in this investigation. Keep your list for review later.

Share your thinking with others in your group and with your class.

Materials Needed

For this investigation your group will need:

- reference resources about weather (books, CD-ROMs, access to the Internet)
- weather instruments (thermometer, wind vane, anemometer, rain gauge)
- instructions for using each weather instrument
- graph paper for making charts
- large flat-bottomed plastic pail or metal can
- street map of town, city, county, or school district
- white, self-adhesive labels, about 2.5 cm × 2.5 cm

 Investigate

Part A: Kinds of Weather Observations

1. Your group members are going to become specialists in making a weather measurement or observation.

INVESTIGATING CLIMATE AND WEATHER

Your job will be to:
- research the science behind your weather observation;
- learn about the techniques for making your observation;
- study any instrument that is needed for your observation;
- practice making your weather observation;
- write a protocol for making your weather observation;
- set up a center for your classmates to learn how to make your weather observation correctly.

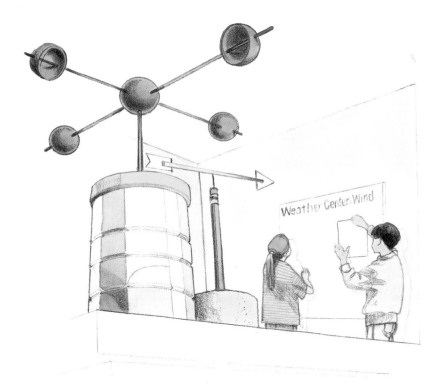

 Check any design with your teacher before attempting to construct a homemade instrument.

2. Your group may be assigned one or more of the following weather observations: temperature, wind speed, wind direction, cloud types, cloud cover, precipitation type, or precipitation amount.

Investigation 1: Observing Weather

Research your weather observation.

a) Describe the science behind the weather observation in your journal.

b) What instrument, if any, is used to make your weather observation?

c) If you did not have a commercial version of your weather instrument, how could you make a homemade version? Draw a sketch of a homemade version in your journal.

3. You will be provided with information on how to make your weather observation properly so that your data are dependable.

Read the information carefully.

a) Write this information in the form of a protocol that others can easily understand and follow. Begin by writing a draft protocol for your weather observation. Remember to include the following in your protocol:

- the technique for taking the data;
- how to locate and set up your instrument properly, if your observation relies upon an instrument;
- how to read your instrument;
- how to record your data;
- the units of measurement to be recorded.

4. Exchange your draft protocol with one from another group in your class.

Read the other group's protocol.

a) Make comments on the other group's protocol. Your comments should be consistent with the criteria for a protocol outlined in Step 3 (a).

5. When you get your protocol back, revise it.

Practice making your weather observation using your protocol. Let each member of your group try this and compare the data you get from each person.

a) Does data vary between group members? If yes, how can the variation be reduced?

Inquiry
Writing a Protocol

A protocol is a procedure for a scientific investigation. It is a set of directions that someone else can read and follow. An important quality that your protocol should have is the ability to be replicated. In other words, your protocol should give consistent and reliable results, for anybody who uses it.

 Check your protocol with your teacher before proceeding to make observations.

INVESTIGATING CLIMATE AND WEATHER

6. Design a center to teach others about your weather observation.

   a) Make a sketch in your journal of how your center will look, and what questions your center will address. Some suggested questions are:
   - What does this weather observation tell you about the weather picture for a particular day?
   - What ideas do you have about how this weather measurement helps in predicting the weather?
   - In thinking about the protocol, what sorts of things could affect the accuracy of the weather data?

   Address any additional questions that you think will help others to understand your center.

7. Construct your center. Set it up where it will work best for visitors to make weather measurements or observations.

8. Work through the centers according to a schedule set by your teacher. Be sure that you are clear on how to follow each protocol. Also be sure that you understand what you are observing.

   a) Record your measurements in your journal.

   b) Write any questions that come to mind as you work through the center.

Part B: Making Observations for a Weather Map

1. Each student in the class will make weather observations at home for a week. The goal is to make a daily weather map of the local region.

   As a class, decide on a protocol for your observations. Here are some of the things you need to think about for the protocol:
   - An ideal weather map shows weather observations from many stations that are uniformly spaced.
   - Observations need to be made at exactly the same time of day.
   - Each observer needs to make exactly the same kinds of observations. Below is a list of these observations, with some helpful comments.

Investigation 1: Observing Weather

Temperature: Your thermometer needs to be mounted in a shady place one to two meters above the ground. Make sure that it will not be blown away by a strong wind. Also make sure it is a shady spot. If you decide not to leave it outside the whole time, give it a few minutes to reach the outdoor temperature after taking it outside. Record temperatures in degrees Fahrenheit and degrees Celsius.

Wind direction: Record the wind direction as north, northeast, east, southeast, south, southwest, west, or northwest. Important: meteorologists record the direction the wind is blowing *from*, not the direction the wind is blowing to. For example, a northeast wind blows from the northeast toward the southwest.

Wind speed: You may not have an anemometer (an instrument that measures the speed of the wind), so you can use the Beaufort scale of wind speed. The Beaufort scale provides an estimate of wind speed based on observed effects of the wind. It is only approximate, but it is useful.

The Beaufort Wind Scale

Beaufort Number	Kilometers per hour	Miles per hour	Wind Name	Land Indication
0	<1	<1	calm	smoke rises vertically
1	1–5	1–3	light air	smoke drifts
2	6–11	4–7	light breeze	leaves rustle
3	12–19	8–12	gentle breeze	small twigs move
4	20–29	13–18	moderate breeze	small branches move
5	30–38	19–24	fresh breeze	small trees sway
6	39–50	25–31	strong breeze	large branches move
7	51–61	32–38	moderate gale	whole trees move
8	62–74	39–46	fresh gale	twigs break off trees
9	75–86	47–54	strong gale	branches break
10	87–101	55–63	whole gale	some trees uprooted
11	102–119	64–73	storm	widespread damage
12	>120	>74	hurricane	severe destruction

INVESTIGATING CLIMATE AND WEATHER

Clouds: Important types of clouds are pictured below. Observe which type or types of clouds are in the sky, and estimate how much of the sky is covered by clouds (zero-tenths, one-tenth, two-tenths, etc., up to complete cloud cover). If you are not sure about the types of clouds, write one or two sentences in your notebook to describe how they look.

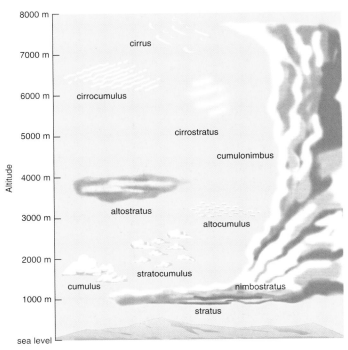

Precipitation: Is rain, drizzle, snow, or sleet falling? If so, is it light, moderate, or heavy? If it has rained since your last observation, how much rain has fallen? If it has snowed since your last observation, what is the depth of new snow? Record rainfall or snowfall in inches and centimeters.

- You need to decide upon a set of symbols for representing your weather observations on the map. The diagram on the following page shows the standard way of doing this on weather maps. Once you get used to this system, the data will show you at a glance what the weather was like at the station.

a) Once everyone has agreed on the protocol, write it down and make copies of it for each student.

b) Make your daily observations, and record them in your journal.

Investigation 1: Observing Weather

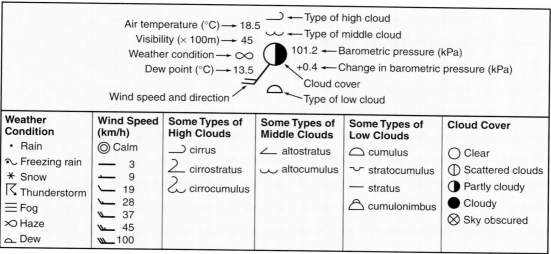

2. Your teacher will supply a base map for making each day's weather map. As a class, mark the locations on the map of all the "stations" that have been decided upon.

   After each day's observations, plot all of the data on the weather map for that day.

3. After all five daily weather maps have been plotted, have a class discussion that deals with the following questions. Record the results of your discussion in your journal.

   a) How much did the weather vary from place to place at the same time over the local region?

   b) Do you think that the variation seen on the maps reflects real variations in the weather, or were they caused by differences in the way observations were made? Explain.

   c) Give reasons for the changes you observed from day to day on your weather maps.

   d) Are there regions of the United States where typical day-to-day variations in the weather are greater than in your local region? Where they are less? What do you think might be the causes for these differences?

Inquiry

Making Observations

Every field of science depends upon observations. Meteorology starts with observations of the weather. It is important that all observations be made carefully and recorded clearly. In this investigation others in your class will be relying on the accuracy of your observations to draw a weather map.

INVESTIGATING CLIMATE AND WEATHER

Digging Deeper

ELEMENTS OF WEATHER

All sciences begin with observations. Without observations, scientists have no way to develop new theories and to test existing theories. The weather is no exception. Meteorologists (scientists who study the weather) observe many elements of the weather. This takes place both at the Earth's surface and at high altitudes. The observations you made in this investigation include many of the most important elements of the weather. Weather observations are needed both for predicting the weather and for developing and testing new theories about how the weather works.

As You Read...
Think about:
1. What does air temperature measure?
2. Why and how does wind blow?
3. How are clouds formed?
4. How is rain formed?

Air Temperature

Air consists of gas molecules, mostly nitrogen (N_2) and oxygen (O_2). Although you cannot see them with your eyes, the molecules are constantly moving this way and that way at very high speeds. As they move, they collide with one another and with solid and liquid surfaces. The temperature of air is a measure of the average motion of the molecules. The more energy of motion the molecules have, the higher the air temperature.

Air temperature is measured with thermometers. Common thermometers consist of a liquid-in-glass tube attached to a scale. The scale can be marked (graduated) in degrees Celsius or degrees Fahrenheit. The tube contains a liquid that is supplied from a reservoir, or "bulb," at the base of the thermometer. Sometimes the liquid is mercury and sometimes it is red-colored alcohol. As the liquid in the bulb is heated, the liquid expands and rises up in the tube. Conversely, as the liquid in the bulb is cooled, the liquid contracts and falls in the tube.

Investigation 1: Observing Weather

When you are measuring the air temperature, be sure to mount the thermometer in the shade. If the Sun shines on the thermometer, it heats the liquid, and the reading is higher than the true air temperature. Also, when you take the thermometer outside, give it enough time to adjust to the outdoor air temperature. That might take several minutes.

Wind

The wind blows because air pressure is higher in one place than in another place. The air moves from areas of higher pressure to areas of lower pressure. Also, the greater the difference in air pressure from one place to another, the stronger the wind. Objects like buildings, trees, and hills affect both the direction and speed of the wind near the surface. To get the best idea of the wind direction, try to stand far away from such objects. A park or a playing field is the best place to observe the wind.

Wind speed is measured with an anemometer. Most anemometers have horizontal shafts arranged like the spokes of a wheel. A cup is attached to the end of each shaft. The wind pushes the concave side of the cup more than the convex side, so the anemometer spins in the wind. The stronger the wind, the faster the cups spin. The cup spin rate is calibrated in terms of wind speed (e.g., miles or kilometers per hour).

INVESTIGATING CLIMATE AND WEATHER

You do not need an anemometer to estimate the wind speed. You can use a verbal scale, called the Beaufort scale (page C5). It describes the effect of the wind on everyday things like trees.

Wind direction is measured with a wind vane. You can also estimate the wind direction by yourself just by using your face as a "sensor." Face into the wind, and then record the direction you are facing.

Clouds

Clouds are formed when humid air rises upward. As the air rises, it expands and becomes colder. With enough cooling, water vapor condenses into tiny water droplets (or deposits into tiny ice crystals). The droplets or crystals are visible as clouds. Condensation is the change from water vapor to liquid water. Deposition is the change from water vapor directly to ice crystals. Condensation or deposition takes place when air is cooled to its dew-point temperature. When humid air is cooled at the ground (that is, when air reaches the dew point at ground level), fog is formed. You will learn more about clouds later in this module.

Clouds form at a wide range of altitudes, from near the ground to very high in the atmosphere. The appearance of clouds varies a lot, depending on the motions of the air as the clouds are formed. Other important things to observe about clouds are the percentage of the sky they cover, where they are located in the sky, how much of the sky they cover, and their direction of movement. A good way to find their direction of movement is to stand under a tree branch or an overhang of a building and watch the clouds move relative to that stationary object. Clouds move with the wind, so observing cloud motion provides information on the wind direction high in the atmosphere.

Investigation 1: Observing Weather

Cirrus

Altostratus

Altocumulus

Nimbostratus

Cumulonimbus

Cumulus

Stratocumulus

Stratus

INVESTIGATING CLIMATE AND WEATHER

Precipitation

Raindrops are formed when the cloud droplets grow large enough to fall out of the clouds. Most of the rain that falls in the winter, and even much of what falls in the summer, is from melting of snowflakes as they fall through warmer air.

Rainfall is measured by the depth of water that falls on a level surface without soaking into the ground. Rainfall is measured with a rain gauge. A basic rain gauge is nothing more than a cylindrical container, like a metal can with a flat bottom, that is open to the sky. The only problem is to get an accurate measurement of the depth of water that has fallen. Accurate rain gauges are designed so that the water that falls into the container is funneled into a much narrower cylinder inside. In that way, the depth of the water is magnified, and is easier to read.

If you live in a part of the United States where it snows in winter, you can easily measure the snow depth with a ruler graduated in either centimeters or inches. The best time to make the measurement is right after the snow stops falling. The measurement can be tricky, because wind can cause snow to drift. The best place to measure snow depth is on level ground far away from buildings and trees. Take measurements at several spots and compute an average.

Investigation 1: Observing Weather

Review and Reflect

Review

1. List the weather instruments you studied in this investigation and describe what each measures.
2. What factors do you need to consider to get an accurate reading of each of the following:
   a) air temperature?
   b) wind speed and direction?
   c) rainfall? snowfall?

Reflect

3. What are some drawbacks to relying on weather measurements taken by your class, as opposed to following the reports issued by professionals?
4. Most of the weather observations you made can be made and recorded automatically by instruments. Using one kind of observation as an example, describe an advantage and disadvantage of using technology to record weather information versus people recording the information.

Thinking about the Earth System

5. How do seasonal changes in air temperature affect plants and animals?
6. How does the wind interact with the surface waters of the ocean or large lake?
7. How does the wind affect trees, bushes, and other vegetation?
8. How does air temperature and wind speed influence how comfortable you are when outside?

Thinking about Scientific Inquiry

9. How is writing and following a protocol important to the inquiry process?
10. How did you use mathematics in this investigation?
11. Describe an example of how you collected and reviewed data using tools.

INVESTIGATING CLIMATE AND WEATHER

Investigation 2:

Comparing Weather Reports

Key Question

Before you begin, first think about this key question.

What does a weather report contain?

Think about what you already know about weather reports. What information is contained in one? What do you need to know to interpret a weather report?

Share your thinking with others in your group and with your class.

Materials Needed

For this investigation, your group will need:

- sample weather report
- weather reports or forecasts for three consecutive days, from one of the following media: television, national newspaper, local newspaper, commercial or public radio, the NOAA weather radio, Internet, telephone
- weather data for each of the three days following the reports (temperature, wind speed, precipitation, cloud cover, humidity, etc.)
- reference resources about weather

Investigate

1. A *weather report* tells you what the weather conditions are at the time. A *weather forecast* tells you what the weather is likely to be in the future. Your group will receive one of several different kinds of weather reports.

   a) Make a list of all the weather words in the report.

Investigation 2: Comparing Weather Reports

b) Turn the list into a table like the one below. If someone in your group knows some of the other information for the table, fill in that part. If not, leave it blank for later. If you find a term that you do not understand, write it down as a question to answer later.

Weather Word	Descriptors Used	Definition	Instrument Used for Measuring	Unit of Measurement
Wind Speed				
Wind Direction				
Clouds				
Kind of Precipitation				
Amount of Precipitation				
Temperature				
Pressure				
Other				

2. Study your group's report a second time.

   a) This time, make a table that shows what kind of weather information your report includes.

   The table should clearly show the kinds of weather reports your class is studying (newspaper, radio, and so on),

INVESTIGATING CLIMATE AND WEATHER

the categories of weather information the report includes (temperature, wind speed, etc.), and the date and time the report was produced.

b) Share your findings with other groups, in order to fill in your table completely.

Weather Report Source	Date & Time of Report	Temperature	Wind Speed	Humidity	Cloud Cover	Other
Local Paper						
National Newspaper						
Radio						
Internet						
Television						
Phone						

Inquiry

Collect and Review Data Using Tools

One important science inquiry skill is the collecting and reviewing of data using tools. In this inquiry you are using tables as data management tools. You could also use a computer as a tool to make your table.

3. Obtain the weather forecasts (temperature, wind speed, precipitation, etc.) from your weather source during a three-day period.

a) Record the forecasts in the form of a table.

b) Record also the actual weather data.

c) Share your findings with other groups, in order to fill in your table completely.

The following is an example of a table that you can use:

Source	Day 1 Forecast	Day 1 Actual Data	Day 2 Forecast	Day 2 Actual Data	Day 3 Forecast	Day 3 Actual Data
Local Paper						
National Newspaper						
Radio						
Internet						
Television						
Phone						

Investigation 2: Comparing Weather Reports

d) Compare the predicted weather with the actual weather, to judge the accuracy of each of the weather forecasts. Look over the information in your table carefully. Rank the weather-forecast sources from most accurate to least accurate. You can do this by giving the number 1 to the most accurate report, 2 to the second, and so on.

e) What evidence do you have to support your rankings?

4. Discuss the following questions within your group. Record the results of your discussion in your journal. (Report in these questions means report or forecast.)

   a) Which kind of weather report has the most information?
   b) How does the format of the report make that possible?
   c) Which kind of weather report has the least information? Explain.
   d) What information do all of the weather reports include?
   e) Why do you think that this information is included in every weather report?
   f) Which kind of weather report best helps you to understand patterns in the weather? Why?
   g) Which weather report would be most helpful for planning an outdoor event three days in the future? Why?
   h) Which weather report would be most helpful if you were traveling to another part of the country? Why?

5. After discussing the weather report questions, share your ideas with the whole class.

   a) What kinds of information need to be included in an accurate and useful weather report?
   b) How do you think you would obtain the information that you need to make a more complete and accurate weather report?

6. Go back to your table of weather terms from the beginning of the investigation.

   Divide up the terms in the table equally among your group members. Use the resources in your classroom, library, and at home to complete the information in the table.

   a) Complete the table in your journal.
   b) Write down any questions you think of as you do your research.

Inquiry

Predictions in Science

Scientists make predictions and justify these with reasons. A meteorologist uses all the data available to make a complete picture of the present and future weather. By using these data, and his or her knowledge of how weather systems form and move, the meteorologist can predict or forecast the weather.

INVESTIGATING CLIMATE AND WEATHER

c) Make a class table of information about weather words by having each group contribute the words it has found. Keep this table posted as a reference for the rest of the investigations in this module. Keep a copy in your journal

As you work through the other investigations in this module, you will probably find other words related to weather and climate. Add these to the table as you accumulate them.

As You Read...
Think about:
1. What is the difference between a weather report and a weather forecast?
2. What information is contained in a detailed weather report?
3. What does normal average temperature mean?
4. How has weather forecasting changed during the past two hundred years?

Digging Deeper

WEATHER REPORTS AND FORECASTS
Weather Reports

Different weather reports contain different amounts of information. The simplest and shortest weather report contains only one piece of information, the present temperature. This kind of report you often hear on the radio. More detailed weather reports contain information about precipitation, wind speed and direction, relative humidity, atmospheric pressure, and so on.

1 TODAY MONDAY, JUNE 11

Chicago: A good deal of sunshine, breezy and warm. Humidity continues slowly upward, and a passing late-afternoon or evening thunderstorm cannot be ruled out. Partly cloudy and mild overnight.

HIGH 84 | LOW 64

A typical weather report tells you the highest and lowest temperatures for the past day. The day's lowest temperature usually occurs just after sunrise. The day's highest temperature is usually reached during early to mid afternoon. A weather report also tells you the present temperature. It may also give you the average temperature for the day. The average daily temperature lies halfway between the highest temperature and the lowest temperature. The weather report might also tell you how

Investigation 2: Comparing Weather Reports

many degrees the average temperature is above or below the normal temperature for that day. The normal temperature is found by averaging the average temperatures for the calendar day for the past 30 years.

Most weather reports give the amount of precipitation (rain or melted snow), if any, that fell during the past day. They also tell you the totals for the current month and the current year. Reports also indicate how much the monthly and annual precipitation totals are above or below normal (the long-term average).

Weather Forecasts

Most people are interested in what the weather will be tomorrow or in the next few days. Predictions of the weather for up to a week in the future are called short-term forecasts. Meteorologists also try to make long-term forecasts (called "outlooks") of the weather for a month, a season, or a whole year. Long-range outlooks are different from short-term forecasts in that they specify expected departures of temperature and precipitation from long-term averages (e.g., colder or warmer than normal, wetter or drier than normal).

In earlier times, before invention of the telegraph and the telephone, weather observations from faraway places could not be collected in one place soon after they were made. In those times, the only way of predicting the weather was to use your local experience. Given the weather on a particular day, what kind of weather usually follows during the next day or two? As you can imagine, the success of such forecasting was not much better than making a random guess.

Beginning in the 1870s, a national weather service used the telegraph to gather weather observations from weather stations located over large areas of the country. Simultaneous weather observations allowed

Investigating Earth Systems

INVESTIGATING CLIMATE AND WEATHER

meteorologists to plot weather maps and follow weather systems as they moved from place to place, greatly improving the accuracy of weather forecasts.

Through the 20th century, meteorologists developed even better tools for observing and predicting the weather. As you will learn in the following investigations, special instruments measure weather in the atmosphere far above the ground. Satellites orbiting the Earth send back images of the weather over broad areas of the planet. In addition, computer models were developed for weather forecasting. The important processes operating in the Earth system that govern weather are built into a computer model. The model starts with the present weather and tries to simulate how the weather will evolve in the future. These computer models are run on supercomputers. They can handle enormous amounts of observational data and make billions of computations quickly. Today's computer models do a very good job of predicting the weather for the next few days. You know, however, that sometimes the forecast is wrong! The science of weather forecasting is still developing.

Weather prediction will never be perfect. One reason is the absence of reliable weather observations from large areas of the globe (especially the oceans). These observations are needed for computer models to accurately represent the present state of the atmosphere. A second reason is that even small changes in the weather in one place can cause much larger changes in weather elsewhere. The effects are small at first, but they become much greater. It's very difficult for computers to simulate these interactions. Although forecasts will never be perfect, they will continue to improve in the years ahead. Through research, meteorologists learn more and more about the details of how weather in the Earth system works.

Investigation 2: Comparing Weather Reports

Review and Reflect

Review

1. What was the most accurate weather forecast that your class studied? Explain.
2. Where do the weather reports and forecasts that your class studied come from?
3. Why does the accuracy of weather forecasts usually decrease as the number of days ahead increases?

Reflect

4. If you were to write your own weather report, what would you include and how would you get your information?
5. Why do you think weather reports vary over a given area?
6. Which kind of weather report is most useful to you on a daily basis? Why is that?

Thinking about the Earth System

7. How has communication technology helped to develop a better understanding of the Earth system? Give an example.
8. Write down all the connections you can think of in this investigation that show a relationship between weather and the Earth's systems. Keep this record on your *Earth System Connection* sheet.

Thinking about Scientific Inquiry

9. What question did you explore in this investigation? Can you answer the question now? Explain.
10. What tools did you use to collect and review data in this investigation?

INVESTIGATING CLIMATE AND WEATHER

Investigation 3:

Weather Maps

Materials Needed

For this part of the investigation your group will need:

- three weather maps from different newspapers
- information on weather-map symbols
- weather map with symbols for clear skies, cloudy skies, partly cloudy skies, and rain
- weather map with symbols for high pressure and low pressure
- overhead transparency sheet
- two or three colors of overhead transparency markers
- weather map with temperatures only
- weather map showing 10° isotherms
- colored pencils

Key Question

Before you begin, first think about this key question.

What can weather maps tell you about weather?

In a previous investigation you learned about some of the techniques used to measure elements of weather. You have already learned some of the ways this information is displayed on maps. What are some of the other ways this is done?

Share your thinking with others in your group and with your class. Keep a record of the discussion in your journal.

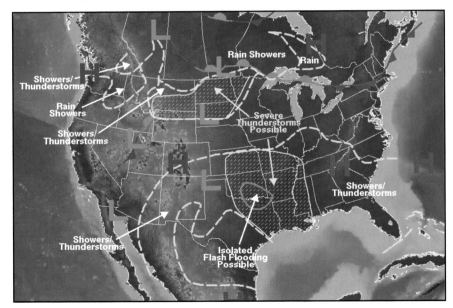

Investigate

Part A: Working with Weather Maps

1. Obtain copies of three newspaper weather maps from different sources.

Investigation 3: Weather Maps

In your group, look for at least five things that the maps have in common.

a) Write these down. Share your list with another group.

b) What kinds of information are included in all three weather maps?

c) Make a list of what you already know about the information on weather maps.

d) Write down what you would like to know.

2. Read any information you have available on weather-map symbols.

Share your information about weather-map symbols with the rest of the class.

a) Make up a class chart of weather symbols to use for reference. Include a copy in your journal.

b) How many of your weather-map questions above can you now answer with this information? Record the answers in your journal.

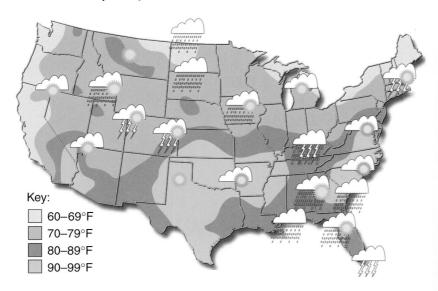

Key:
☐ 60–69°F
☐ 70–79°F
☐ 80–89°F
☐ 90–99°F

3. Obtain a weather map that has symbols for sky conditions on it.

Place an overhead transparency sheet on top of the map and copy the outline of the map.

INVESTIGATING CLIMATE AND WEATHER

Inquiry
Consider Evidence

Scientists look for likely explanations by studying patterns and relationships within evidence. By circling similar types of weather, you developed groupings that reveal patterns. Meteorologists also group different types of weather to reveal patterns that help them correlate factors like high-pressure systems and fair weather.

On the transparency, circle all the areas that have a sunny symbol. Shade these areas with one color of overhead transparency marker.

Make a key showing which color you are using for sunny areas on the map.

4. Repeat this process using another color for precipitation areas. By convention, green is usually used to indicate precipitation on weather maps. Remember that precipitation includes rain, drizzle, snow, sleet, and hail.

   Add this color to your key.

   All the uncolored areas will be cloudy or partly cloudy. You can leave these uncolored or use a new color to show them.

5. Put your transparency sheet on a map that shows high-pressure areas, low-pressure areas, and fronts.

   a) What relationships can you see between the first map, showing the sky conditions and the second map, showing the pressure systems and fronts? Be sure that you are comfortable with the relationships between sky conditions and pressure systems before you move on. Ask for help as you need it.

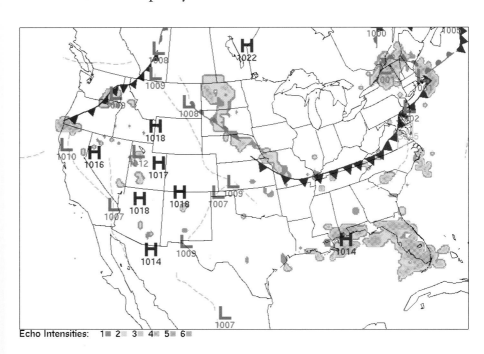

Investigating Earth Systems

Investigation 3: Weather Maps

6. Obtain a weather map showing temperatures only.

   With different colored pencils, draw smoothly curving lines to connect points on the map where temperatures are the same. These curves are called isotherms (*iso* means same, and *therm* stands for temperature). Use a contour interval of 10°F.

   For example, your map might show an isotherm for 50°F. The curve would pass through points where the temperature was 50°F. Drawing contours is not easy, because most of the temperatures shown on the map are different from the ones you selected for the isotherms. You have to interpolate. To interpolate means to find a value that falls in between two other values. For example, suppose that one point on your map has a temperature of 47°F and a nearby point has a temperature of 55°F. You know that the 50°F isotherm has to run somewhere between those two points. Also, it has to be closer to the 47°F point than the 55°F point. That's because 50° is closer to 47° than to 55°. An example has been provided.

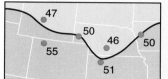

INVESTIGATING CLIMATE AND WEATHER

Draw your isotherms at first with light pencil lines, and use an eraser to adjust them if necessary. Once you feel comfortable with your isotherms, draw them permanently with the colored pencils.

7. Now look at a new weather map that just shows bands of temperature readings. The boundaries between these bands are isotherms.

   Study the temperature bands on your map carefully. Discuss and answer the following questions:

   a) How do the temperature bands help you to understand what weather is like in different parts of the country?

   b) How might this be useful in forecasting the weather conditions for a particular area?

   c) What patterns can you see in temperatures? (Where is it typically warmer, for example? Where are the temperatures typically cooler?)

   d) What reasons can you think of to explain the temperature patterns you notice?

Inquiry

Hypotheses

When you make a prediction and give your reasons for that prediction, you are forming a hypothesis. A hypothesis is a statement of the expected outcome of an experiment or observation, along with an explanation of why this will happen.

A hypothesis is never a guess. It is based on what the scientist already knows about something. A hypothesis is used to design an experiment or observation to find out more about a scientific idea or question.

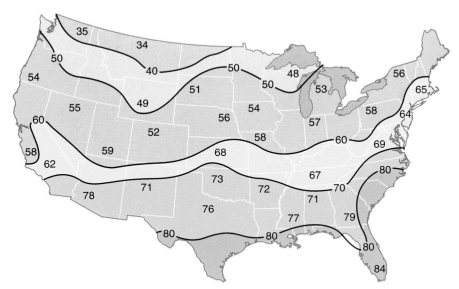

8. Test your knowledge of what a weather map tells you by using a weather map (past or current) to write a weather report for your part of the country.

   a) In your journal write the report so that it is easily understandable by others.

Investigation 3: Weather Maps

Part B: The Movement of Air Masses

1. Read the steps for Part B of the investigation. Although you will use water instead of air, the investigation is a model of what happens when two air masses of different temperature meet.

   a) Before you conduct the investigation, make and record your prediction about what will happen when the card is removed between the two bottles. Also record your reason for your prediction. This is your hypothesis.

2. Fill a 500 mL bottle with cold water and add a few drops of blue food coloring to it.

3. Fill another 500 mL bottle with hot water and add a few drops of red food coloring to it.

   Cap this bottle and shake gently to mix.

   Uncap both bottles.

4. Place a piece of poster board card on top of the cold water bottle.

   Hold the bottle over the sink or the large pan.

   Hold the card tightly to the neck of the bottle, and quickly invert it over the hot water bottle.

   Put the bottles together at their necks. Make sure they match exactly.

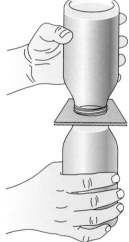

 The water should not be so hot that it can cause burns. Clean up spills.

5. While holding both bottle necks together (cold on top), have someone in your group quickly pull the card out from between the bottles.

   Keep the two necks together and observe what happens to the red and blue water.

   a) Record what you observe.

6. When all water motions seem to be completed in the bottles, empty out the colored water and rinse the bottles for the next group.

Materials Needed

For this part of the investigation your group will need:

- two identical, clear, heavy plastic, 500 mL bottles with medium-size necks
- caps for the two bottles
- one piece of poster board, 10 cm × 10 cm
- supply of hot water and cold water
- blue and red food coloring
- sink or large pan over which the bottles can be inverted

INVESTIGATING CLIMATE AND WEATHER

Answer the following questions in your journal:

a) What happened when the two bottles were put together? How could you explain this?

b) Air is a fluid, just like water. How does what you observed with the two bottles explain what happens when cold air masses and warm air masses meet?

c) When cold and warm air masses meet, what do you expect to happen? Why is that?

d) How did your results compare with your hypothesis?

Part C: Atmospheric Pressure

1. Your teacher will arrange a field trip to a nearby building that is at least four stories tall and has an elevator.

   At the lowest floor, before you get in the elevator, read the pressure with the barometer. To do that, tap the glass front very gently with your fingernail several times and watch the position of the needle on the dial.

   a) Record the average position of the needle as you tap the glass.

Materials Needed

For this part of the investigation your group will need:

- aneroid barometer, with scale marked in inches of mercury

 Stay with an adult supervisor at all times.

Investigation 3: Weather Maps

2. Take the elevator to the top floor, carrying the barometer with you.

   While the elevator is going up, watch the needle. To help the needle adjust its position, you can tap the glass front gently with your fingernail now and then.

3. When the elevator has reached the top floor, step out of the elevator and read the barometer again.

   a) Record how barometric pressure changed as you were riding up the elevator.

   b) Record the barometer reading on the highest floor.

4. Ride the elevator back down to the lowest floor.

   Read the barometer again.

   a) Record this reading, and any relationships between the top and bottom floor readings.

   b) Explain what you think caused the change in the reading of the barometer as you rode up in the elevator.

5. Most barometers record the atmospheric pressure in "inches of mercury." The average atmospheric pressure at sea level (zero elevation of the land surface) is about 30 in.

   Calculate the percentage change in the atmospheric pressure as you went up in your elevator ride.

   a) Find the difference in pressure between the lowest floor and the highest floor.

   b) Divide the difference in pressure by the value for the reading at the bottom floor.

   c) Then multiply the result by 100 to convert your answer to a percentage. This is approximately the percentage of the Earth's atmosphere you went up through on your elevator ride!

6. Look at a weather map that shows barometric pressure across the United States.

   a) Where is the air pressure the lowest? Where is it highest?

   b) What do the H (or High) and L (or Low) symbols indicate on the map?

   c) How does the pressure change moving away from the H? How does the pressure change moving toward the L?

Inquiry
Using Mathematics

In this investigation you made measurements using inches of mercury as a unit to collect data. (In the International System of Units, pressure is measured in pascals, symbol Pa.) Calculations were then used to interpret the data you collected. Scientists also use mathematics in their investigations.

INVESTIGATING CLIMATE AND WEATHER

d) What is the general relationship between air pressure (high or low) and the type of weather (storms, sunny, etc.) in a region?

e) In a rapidly ascending elevator or an airplane taking off, you may feel a popping sensation in your ears. What do you suppose causes this, and how might it be related to the drop of air pressure with increasing altitude?

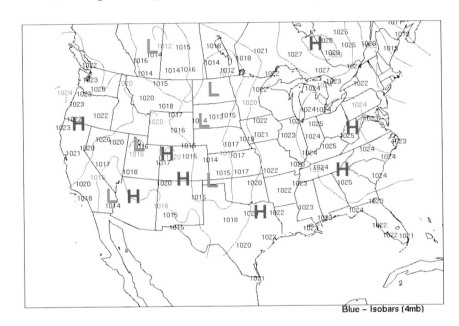

Blue – Isobars (4mb)

As You Read...
Think about:
1. Who drew the first weather map, and how was the information obtained?
2. What is an isotherm? What is an isobar?
3. Why does air pressure always decrease upward in the atmosphere?
4. Why is the weather in high-pressure areas usually fair? Why is the weather in low-pressure areas usually cloudy and stormy?
5. What is the difference between a cold front and a warm front?

Digging Deeper

WEATHER MAPS

A weather map is a graphical model of the state of the atmosphere over a broad region at a specific time. Meteorologists call these synoptic maps (*syn-* means at the same time, and *-optic* stands for seeing). In the late 1700s, Benjamin Franklin drew the first synoptic weather map. He asked a number of friends living in eastern North America to record the weather each day for several days and then mail their journals to him. He then drew weather maps for each day. The mails were very

Investigating Earth Systems

Investigation 3: Weather Maps

slow then, so that Franklin was not able to draw the maps until long after the observations were made. Nonetheless, the maps were useful because they revealed, for the first time, that winds blow around the centers of storm systems that move day by day.

Weather maps that are available to the public vary a lot in how much information they show. Most weather maps in newspapers and on television show only temperature bands that are defined by isotherms, areas of precipitation, the location of high-pressure systems and low-pressure systems, and weather fronts. More specialized weather maps show air pressure, wind speed and direction, cloud cover, and precipitation. The maps you created in Investigation 1 were a simplified version of this kind of weather map.

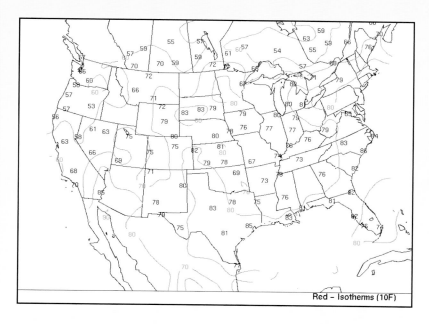

Atmospheric Pressure

Air has weight. That idea might seem strange to you, because air seems very thin, even at sea level. Remember, however, that the atmosphere extends to great

INVESTIGATING CLIMATE AND WEATHER

altitudes. You can think of air pressure as the weight of a column of air above a unit area on the Earth's surface. The column of air above a square area that is one inch on a side averages about 14.7 lb. at sea level. In the metric system, that's about 1.0 kg/cm^2 (square centimeter). If you try to pump the air out of a closed container, the outside air pressure will cause the container to collapse unless the container is very strong. The reason you don't feel air pressure is that the pressure inside your body is equal to the air pressure acting on the outside of your body!

You saw from your elevator ride with the barometer that the air pressure decreases upward in the atmosphere. That's because at higher levels in the atmosphere there is less air above to cause the pressure.

Detailed weather maps show the atmospheric pressure by means of curved lines called isobars. As with an isotherm for temperature, an isobar connects all points with the same atmospheric pressure. Air pressure always decreases with altitude so that the pressure at the land surface is less where the elevation of the surface is high. To remove the effect of elevation on air pressure readings, meteorologists adjust the readings to what they would be if the weather station were at sea level. The adjusted pressure is what you would measure if you could dig a very deep shaft all the way down to sea level and put your barometer at the bottom of the shaft. The adjusted pressure is plotted on weather maps.

High-Pressure Areas and Low-Pressure Areas

Most weather maps show areas, labeled with an **H** (or **High**), where the atmospheric pressure is relatively high, and areas labeled with an **L** (or **Low**) where the atmospheric pressure is relatively low. The isobars around such areas are usually closed curves with the approximate shape of circles. Viewed from above, surface winds blow clockwise (in the Northern Hemisphere) and outward

Investigation 3: Weather Maps

about the center of a high as shown in the diagram. As air leaves the high-pressure area, the remaining air sinks slowly downward to take its place. That makes clouds and precipitation scarce, because clouds depend on rising air for condensation of water vapor. High-pressure areas usually are areas of fair, settled weather. Viewed from above, surface winds blow counterclockwise (in the Northern Hemisphere) and inward about the center of a low as shown in the diagram. Converging surface winds cause air to rise, producing clouds. Low-pressure areas tend to be stormy weather systems.

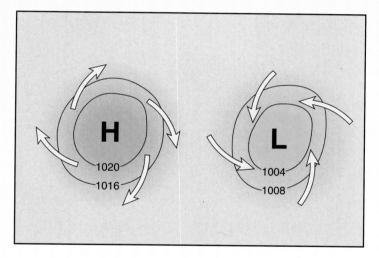

Air Masses and Fronts

Large masses of air, as much as 1000 km across, take on certain weather characteristics when they stay at high latitudes (near the poles) or at low latitudes (near the Equator) for weeks at a time. They may be very cold or very warm, or they may be very humid or very dry. Then, as they move into other areas, they can cause changes in the weather. The coldest winter weather in much of the United States occurs when a bitter cold air mass from the high arctic regions of northeastern Asia, Alaska, or northern Canada sweeps down into the southern parts of North America. At other times, a flow of warm ➔

INVESTIGATING CLIMATE AND WEATHER

and humid air from the tropics causes uncomfortably muggy weather over the eastern United States.

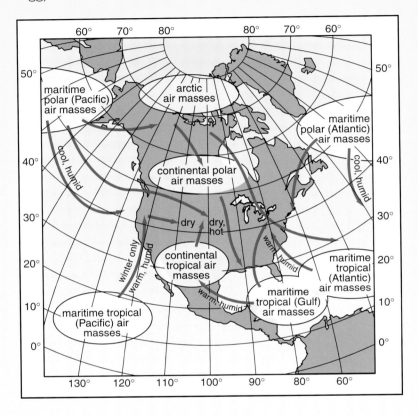

The boundaries between air masses are often zones of very rapid changes in temperature and humidity. Storms (low-pressure systems) tend to develop along these zones of rapid change. The line with the triangular teeth, called a cold front, shows where the cold air mass is wedging under the warm air mass. (See the cross-sectional diagram on the following page.) As the warm air is lifted along the front, thunderstorms may develop. The line in the diagram with the circular teeth, called a warm front, shows where the warm air mass is moving up over the cold air mass. Broad areas of light to moderate rain or snow are often associated with the warm front.

Investigation 3: Weather Maps

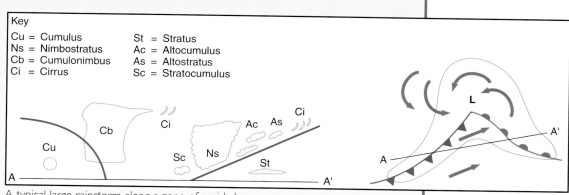

A typical large rainstorm along a zone of rapid change.

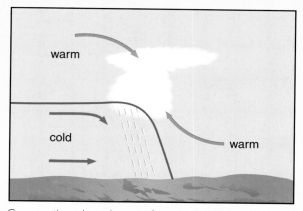
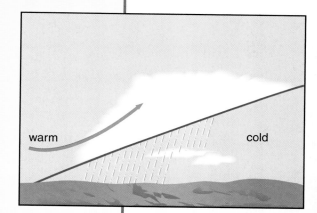

Cross-sections through warm fronts and cold fronts.

If the pattern of the fronts in the diagram looks to you a bit like a wave, you are right. A low-pressure system is developing like a wave along the boundary between the two air masses! You can see that the pattern of surface winds, shown by the arrows, is developing a counterclockwise and inward pattern. After a storm like this develops, the swirling pattern of winds continues for many days, until it finally dies out. A storm of this kind can move hundreds of kilometers in a day, or it can remain in almost the same position for a day or even longer. Much like a Frisbee, the storm combines rotational and translational motion as it travels across the nation.

INVESTIGATING CLIMATE AND WEATHER

Review and Reflect

Review

1. What kinds of information were plotted on the weather maps that you investigated?
2. Use your observations of the two bottles of water in Part B of this investigation to explain what happens when a cold air mass meets a warm air mass. Which air mass is likely to rise? Which air mass is likely to stay near Earth's surface?
3. What evidence did you obtain from your investigation that shows that air pressure decreases with altitude?

Reflect

4. What do weather maps tell you about the weather now and in the future?
5. What evidence of relationships were you able to find between cloud patterns and precipitation?
6. What evidence of relationships were you able to find between cloud patterns and high- and low-pressure systems?

Thinking about the Earth System

7. On your *Earth System Connection* sheet, note how the things you learned in this investigation connect to the different Earth systems.

    a) Describe how air pressure (atmosphere) is related to elevation (geosphere).

    b) Describe how air masses (atmosphere) are related to the regions over which they form (geosphere or hydrosphere).

Thinking about Scientific Inquiry

8. What weather data did you use to look for relationships in this investigation?
9. How did you use mathematics in this investigation?
10. Give an example of how you used evidence to develop ideas in your investigations into weather.

Investigation 4:
Weather Radiosondes, Satellites, and Radar

Key Question
Before you begin, first think about this key question.

In addition to weather observations on the Earth's surface, how else is weather data collected?

Materials Needed

For this investigation your group will need:

- two pieces of graph paper

In the previous investigation, you studied weather maps. Some of the information used to make the maps is obtained at weather stations at the surface. Much of the information, however, is obtained from other sources. How else can weather observations be made?

Share your thinking with others in your group and with your class. Keep a record of the discussion in your journal.

Investigate

Part A: Data from Radiosondes

1. The table on the following page gives data on how temperature changes with altitude.

INVESTIGATING CLIMATE AND WEATHER

Radiosonde Data June 26, 2001

Jacksonville, Florida		Fairbanks, Alaska	
Temperature (°C)	Altitude (m)	Temperature (°C)	Altitude (m)
20.6	9	17	138
24.2	88	16.4	197
24.6	203	14.1	610
24.8	327	12.8	842
23.2	610	9.7	1219
21.6	884	7	1545
18.4	1219	3	2164
13.2	1829	−2.5	2923
11.2	2134	−6.4	3658
5	3224	−11.4	4540
−0.3	4267	−16.3	5180
−4.3	4997	−22.1	5970
−12.7	6096	−27.7	6746
−23.7	7570	−32.7	7620
−28.3	8230	−38.7	8469
−39.1	9610	−44.9	9300
−43.7	10830	−54.7	10668
−56.7	11983	−57.7	11900
−59.7	13766	−51.7	12719
−64.1	15240	−50.3	15240
−64.9	16540	−49.3	16619
−67.7	17692	−50.1	18601
−64.9	18700	−48.1	20950
−59.7	20780	−48.6	22555
−54.2	22555	−46.6	23774
−49.8	26518	−43.6	25603
−50.7	27432	−40	27432
−42.1	30480	−36.3	29403
−39.9	31270	−33.3	31840
−39	33223	−28.1	33528

A radiosonde is an instrument package that is carried upward by a balloon. As it rises to great altitudes it makes weather observations.

Investigation 4: Weather Radiosondes, Satellites, and Radar

2. For each data set, use a sheet of graph paper to plot how temperature changes with altitude.

   a) Use the vertical axis for altitude above sea level. Use the horizontal axis for temperature. See the sample shown.

   b) Plot the temperature at each altitude.

   c) Then connect the points (which scientists call "data points") with a continuous line. A line like this is called a sounding.

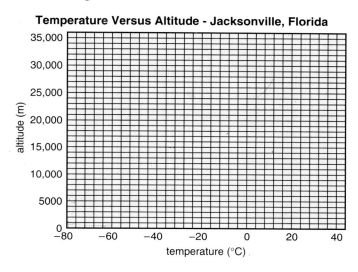

3. Use your graph to answer the following questions:

   a) Does the air temperature generally increase or decrease with altitude?

   b) What do you think is the cause of the increase or decrease?

   c) The "cruising altitude" of commercial jetliners is usually in the range of 10,000 m (about 30,000 ft.) to 13,000 m (about 40,000 ft.). Judging from the two temperature profiles you plotted, what is the typical air temperature in that range of altitudes?

   d) In the Jacksonville sounding, the temperature shows an increase with altitude in the lower part of the atmosphere. What do you think is the cause of that increase? (Hint: the lower atmosphere is heated and cooled from below.)

Inquiry

Mathematical Relationships

In the previous investigation you discovered that the higher the altitude, the lower the air pressure. This is an example of an inverse relationship. In this investigation you are studying another inverse relationship.

Investigating Earth Systems

Part B: Satellite Images

1. Look carefully at the satellite image.
    a) What do you think it shows? How can you check your ideas?
    b) How might satellite images be helpful in forecasting or understanding weather? What do you think?
    c) Compare your ideas with those from another group. How do they compare?

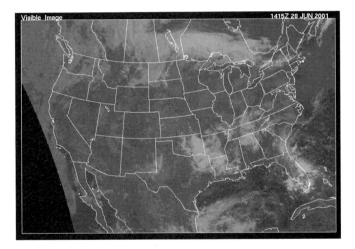

2. Compare the satellite image to the sky conditions shown on a weather map for the same time period.
    a) List and describe as many relationships as you can between the satellite image and the weather elements on your map. Discuss the relationships you discovered with another group.

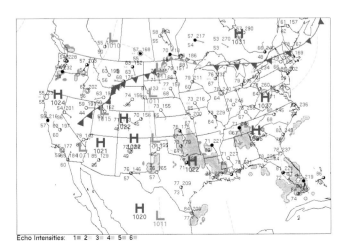

Investigation 4: Weather Radiosondes, Satellites, and Radar

3. Over the next few days, visit a web site that has both satellite images and weather maps. You might also watch television weather reports that show both weather maps and satellite images.

   a) Do you see any pattern in the way weather systems move? Explain.

   b) Compare the movement of weather systems and the map of the movement of air masses (page C34). What relationships can you see?

Inquiry

Using Evidence Collected by Others

In this investigation you used evidence that you were provided to formulate your ideas about weather systems. Meteorologists must also consider the evidence provided by others to develop their ideas about weather patterns.

Part C: Radar Images

1. Compare the radar and satellite images shown on the following page.

   a) What do you notice about the relationship between the clouds and radar echoes?

   b) Are all the clouds producing precipitation? How can you tell?

INVESTIGATING CLIMATE AND WEATHER

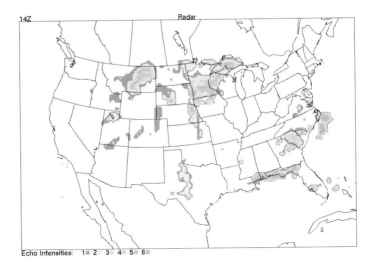

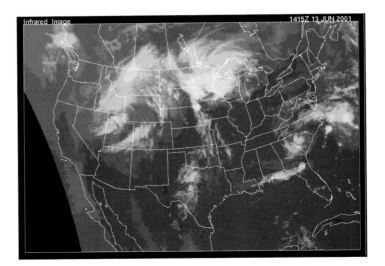

2. Look at the radar image at the top of the next page. It was taken four hours later.

   a) What has happened to the area coverage and intensity of the precipitation over Minnesota and Wisconsin?

   b) What has happened to the area coverage and intensity of the precipitation over North and South Carolina?

3. Look at the infrared satellite image on the next page taken four hours later.

   a) What has happened to the clouds along the border between West Virginia and Virginia?

   b) What do you think is responsible for this change?

Investigation 4: Weather Radiosondes, Satellites, and Radar

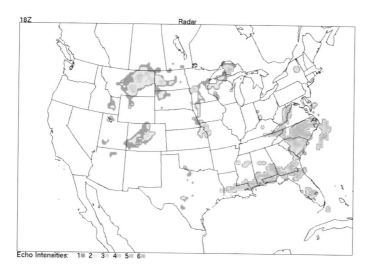

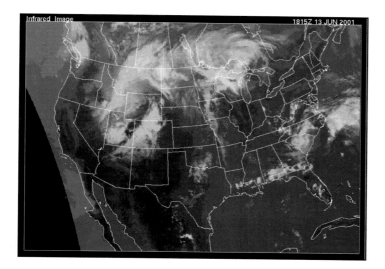

4. Note that on the infrared satellite image, temperature is given on a gray scale—ranging from bright white for lowest temperatures (high clouds) to black for highest temperatures (land). The temperature decreases with altitude, so that high clouds are cold and appear brighter than warmer low clouds.

a) What do the bright white in the clouds over West Virginia and Virginia indicate?

b) Thunderstorm cloud tops rise to great heights. What kind of weather do you think that this area of West Virginia and Virginia was having at the time?

INVESTIGATING CLIMATE AND WEATHER

As You Read...
Think about:
1. Why does temperature usually decrease with altitude in the lower portion of the atmosphere?
2. What weather observations are typically made by radiosondes?
3. What piece of weather information are satellites especially good at showing?
4. How is information from radiosondes and satellites transmitted to the Earth?

Digging Deeper

THE WEATHER HIGH IN THE ATMOSPHERE
Radiosondes

You experience weather at the Earth's surface, but there is weather high in the atmosphere also. What is the weather like at high altitudes? Have you ever taken a ride in a hot-air balloon or climbed a mountain? You would know that the air temperature usually decreases with altitude. The basic reason has to do with how the atmosphere receives and loses its heat energy. The Earth's surface is heated by the Sun at some times and places, that is, when and where it is daylight. It loses heat to outer space at all times (day and night) everywhere. On a global and average annual basis, however, the Earth's surface gains more heat than it loses. The atmosphere near the ground is then heated by the ground. High up in the atmosphere, however, the air loses more heat to space than it absorbs from sunlight. Hence, air usually gets colder with increasing altitude—at least up to an altitude of about 10,000 m (33,000 ft.).

Investigation 4: Weather Radiosondes, Satellites, and Radar

How is the weather in the upper atmosphere measured? It is difficult and expensive to measure the weather in the upper atmosphere. But knowing what conditions are like in the upper atmosphere is important for modern weather forecasting. Since the late 1930s, meteorologists have relied mainly on radiosondes for profiles of temperature, pressure, and humidity from Earth's surface to the upper atmosphere (to altitudes of 30,000 m or 100,000 ft.). Balloons carry radiosondes up through the atmosphere. As they rise, they send back measurements by radio. Tracking of radiosonde movements from the ground shows wind speed and direction in the upper atmosphere. Eventually the balloon bursts, and the instruments fall back to Earth by parachute. (Some radiosondes are recovered and reused.) Their fall is not dangerous to humans, because the instruments are very small and light. Radiosondes are launched at the same time every 12 hours at hundreds of weather stations around the world. A similar instrument, called a dropwindsonde, is released from an aircraft to determine atmospheric conditions in areas where radiosonde data is absent (e.g., in a hurricane over the ocean).

Weather Radar

Radar (**RA**dio **D**etection **A**nd **R**anging) has become an essential tool for observing and predicting weather. Radar was invented and developed in Britain and the United States at the beginning of World War II. It was used to detect the approach of enemy airplanes. An antenna

INVESTIGATING CLIMATE AND WEATHER

sends out pulses of microwave energy. These waves are reflected from solid or liquid precipitation particles in the air and received back by the antenna. The radar equipment shows the position and distance of the particles. The results (the radar echoes) are shown as blotches on a screen, similar to a television or computer monitor. Radar echoes are electronically superimposed on a map of the area to show the location of areas of precipitation. The strength of the echoes is used to find the intensity of precipitation and to tell frozen forms (e.g., hail) from unfrozen forms of precipitation (e.g., rain). Meteorologists use radar to track the movement and follow the development of storm systems, especially small-scale systems like thunderstorms.

Weather Satellites

The first weather satellite was launched into orbit on April 1, 1960, when the United States orbited TIROS-1. Today, satellites are a routine and valuable tool in monitoring the Earth system. Satellite sensors obtain images of the Earth's weather from space. They are especially good at showing cloud cover and the temperatures of clouds and other surfaces in the sensors' field of view. A visible satellite image is like a black-and-white photograph of the planet and is available only for the areas of the planet that are in sunlight. An infrared satellite image shows the temperature of surfaces based on the invisible infrared (heat) radiation emitted by objects. Infrared satellite imagery is available both day and night and is the usual satellite picture shown on televised weathercasts. The most useful satellites are ones in a geostationary orbit. That is an orbit adjusted so that the speed and direction of the satellite matches the Earth's rotation. Then the satellite is always above the same point on Earth's surface and views the same area of the planet.

Investigation 4: Weather Radiosondes, Satellites, and Radar

Review and Reflect

Review

1. How do meteorologists make observations of the weather at high altitudes without going there in airplanes?
2. How are radiosondes and weather satellites similar? How are they different?
3. How is radar used to track thunderstorms?

Reflect

4. What advantages are there in looking at the world from space to observe weather?
5. In what ways, other than those you investigated, can meteorologists obtain weather data from the upper atmosphere?

Thinking about the Earth System

6. How might remote sensing by satellite be used to monitor the geosphere and hydrosphere?
7. How might satellites be used to observe features of the landscape like the distribution of vegetation and ice and snow cover?

Thinking about Scientific Inquiry

8. Give three examples of inverse relationships related to weather that you have discovered in your investigations.
9. How did you use evidence to develop ideas in this investigation?

INVESTIGATING CLIMATE AND WEATHER

Investigation 5:
The Causes of Weather

 Key Question
Before you begin, first think about this key question.

What causes weather?

Materials Needed

For each station in this investigation you will need:

- paper towels
- student journal

Think about what you know about weather patterns and weather reports from previous investigations. How does weather originate?

Share your thinking with others in your class. Keep a record of the discussion in your journal.

Investigate

Part A: Visiting the Stations

1. There will be a series of stations for you to visit. The investigations at the stations will allow you to ask and answer questions about:

Investigation 5: The Causes of Weather

Station 1: The Effects of the Wind
Station 2: Cloud Formation
Station 3: Temperature and Air Pressure

Keep a record of what you do and discover at each of the stations. At the end of your "station journey," you will be looking for common threads that seem to be a part of all aspects of weather.

Station 1: The Effects of the Wind

1. Wet the back of one of your hands with room-temperature water. Leave the other hand dry.
2. Have someone turn on the fan and direct the wind toward the backs of both your hands at the same time.

   a) Observe and record any differences in how your hands feel.

   b) Explain any differences you feel.

   c) How does this experience help to explain why winds help to cool you down on a hot day? Why a cold day feels even colder when the wind is blowing?

Materials Needed

For this station your group will need:

- water supply
- battery-powered fan
- alcohol thermometers
- tape
- cotton batting

INVESTIGATING CLIMATE AND WEATHER

Inquiry
Models in Scientific Inquiry

In this investigation you are using models. A model is the approximate representation (or simulation) of a real system. For example, a weather map is a graphical model of the state of the atmosphere over a given area. Models are used by scientists to study processes that happen too slow, too quick, or on too small a scale to be observed; that are too vast to be changed deliberately; or that might be dangerous. Models are also used to organize your thinking on some complex process. These are called conceptual models.

Be careful that the fan does not touch the thermometers.

3. Using your results, develop a hypothesis related to the rest of this investigation.

   a) What do you predict will happen? Why?

4. Two thermometers are taped to a cardboard stand about 50 cm apart.

5. Read the temperatures on both thermometers

   a) Record these temperatures.

6. Hold a fan about 10 cm away from the bulb of one thermometer and turn on the blades. Observe what happens to both thermometers.

   a) Record your observations.

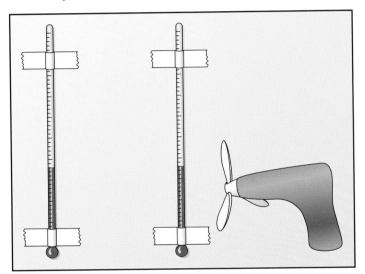

7. Next, dampen a small amount of cotton batting and tape it to one of the thermometer bulbs, so that it is exposed to the air.

   Direct a fan at the bulb of both the wrapped and the unwrapped thermometers.

   Observe what happens to the temperature readings on both thermometers.

   a) Record your observations.

   b) Compare your observations when the thermometer bulb was not dampened and when the bulb was dampened.

   c) Review your hypothesis. Explain why you think there was a difference in the temperatures of the two thermometers.

Investigation 5: The Causes of Weather

Station 2: Cloud Formation

1. Place a metal container in a cooler.

   Put a brick in the bottom of the metal container to weigh it down.

   Place the lid on the metal container.

   Pack ice around the metal container in the cooler.

   Put the lid on the cooler.

2. After fifteen minutes or so the air in the metal container has been chilled. Carefully remove the lid of the cooler.

   Slowly raise the lid of the metal container a few centimeters above its rim.

   Shine a beam from a flashlight into the metal container.

   Take a deep breath; hold it for a few seconds.

   Put your mouth close to the top of the container. Very slowly and gently breathe some air out through your mouth down into the container.

   a) Record your observations.

   b) Describe how what you observed relates to the formation of clouds.

   c) Explain how you think clouds are formed.

Materials Needed

For this station your group will need:

- Styrofoam® picnic cooler
- brick or other heavy mass
- metal container, with lid, small enough to fit in the cooler but large enough to contain the brick
- bag of ice cubes
- flashlight

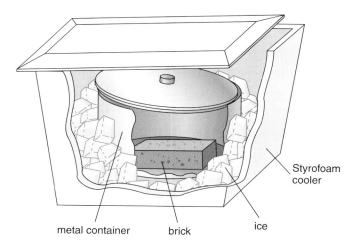

Clean up spills immediately.

3. Using the flashlight beam, try especially to observe the size of the particles in the cloud.

   a) Record your observations.

INVESTIGATING CLIMATE AND WEATHER

Station 3: Temperature and Air Pressure

Materials Needed

For this station your group will need:

- two large round balloons (same size)
- metric measuring tape (or string and meter stick)
- two thermometers
- ice bath

1. Blow up two balloons to the same size and tie them shut.

   a) What do you think will happen when one balloon is cooled and the other is not? Why?

2. Measure the circumference (distance around the center) of both balloons.

   a) Record your measurements.

3. Read the temperature of two thermometers. They should both be at the same room temperature.

   a) Record this temperature.

4. Put one balloon and one thermometer into an ice bath for 10 min. Let the other balloon and thermometer stay at room temperature.

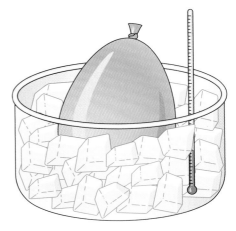

5. After 10 min, read the temperature of the thermometer in the ice bath and the temperature of the room.

   Take the balloon and thermometer out of the ice bath.

   Immediately measure the circumference of each balloon.

   a) Record the temperatures and the circumferences in your journal.

   b) What happened to the balloon in the ice bath?

   c) Did any air escape from the balloon in the ice bath?

   d) Did the density of the balloon increase or decrease because of cooling?

   e) What do you think was happening to the air inside the balloon?

Clean up spills immediately. Dry the thermometer so it is not slippery.

Investigation 5: The Causes of Weather

6. Let both balloons stay at room temperature for five minutes.

   Observe both balloons over that time.

   Measure the circumference of each balloon again after five minutes.

   a) Record your observations and your measurements.

   b) What can this investigation tell you about warm and cold air masses? Which air mass would you expect to be denser: warm or cool? Which type of air mass might take up more space: warm or cool?

   c) If a warm and a cool air mass were to meet, what do you think would happen, and why?

   d) How is air pressure involved?

Part B: Sharing and Discussing Your Findings

1. When your group has completed all of the stations, share your findings with one other group in the class.

   Help each other to answer questions you might have about the science behind the weather. Look for areas that are common across the stations.

   a) In your journal record anything new that you discover during your discussion.

2. When you have finished, hold a class discussion about the science behind the weather.

   Make one large list of the common science elements for the whole class.

   a) Keep a record of the list in your student journal.

Inquiry

Sharing Findings

An important part of a scientific experiment is sharing the results with others. Scientists do this whenever they think that they have discovered scientifically interesting and important information that other scientists might want to know about. This is called disseminating research findings. In this investigation you are sharing your findings with other groups.

INVESTIGATING CLIMATE AND WEATHER

As You Read...
Think about:
1. What is the water cycle?
2. What is the difference between evaporation and condensation?
3. On the Earth's surface, where would you likely find more evaporation than condensation occurring? Where would you likely find more condensation than evaporation?
4. How are clouds formed?
5. How is rain formed?
6. How does temperature affect air pressure?

Digging Deeper

THE WATER CYCLE
The Main Loop of the Water Cycle

A "closed system" consists of a container that allows energy, but not matter, to pass back and forth across the walls of the container. The Earth's atmosphere, ocean, and land surface act as an almost closed system. Water moves along a variety of pathways in this closed system. This system of movement is the global water cycle.

There is one main loop in the water cycle. Water evaporates at the Earth's surface and then moves as water vapor into the atmosphere. The water vapor condenses into clouds and falls from clouds as precipitation back to Earth's surface. Water that falls onto the continent can follow a number of pathways: some water evaporates back into the atmosphere, some is temporarily stored in lakes, reservoirs, and glaciers, some seeps into the ground as soil moisture and ground water, and some runs off into rivers and streams. Ultimately, all the water on land drains into the ocean.

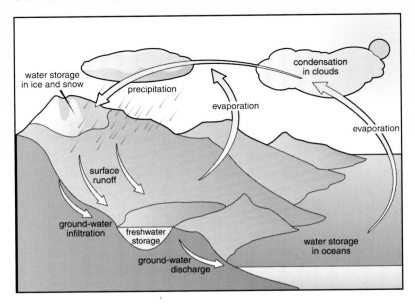

Investigation 5: The Causes of Weather

Each year, there is an excess of evaporation over precipitation on the oceans, and an excess of precipitation over evaporation on the continents. The net gain of water on the continents equals the net loss of water from the ocean. Under the influence of gravity, the excess water on land flows to the sea.

Evaporation and Condensation

With the range of temperature and pressure conditions in the Earth system, water coexists in all three phases (solid, liquid, vapor). It is continually changing from one phase to another. In the solid phase, water molecules vibrate about fixed locations, so an ice cube is crystalline and retains its shape (as long as its temperature is below freezing). In the liquid phase, water molecules have considerably more energy. The molecules are free to move around one another. For this reason, water takes the shape of its container. In the vapor phase, water molecules have the most energy. Even a small amount of water vapor spreads evenly throughout the volume of any container.

At the interface between water and air (e.g., the surface of the ocean or lake), water molecules move in two directions. Some molecules leave the water surface to become vapor, and some molecules leave the vapor phase to become liquid. Evaporation occurs if the flux of water molecules becoming vapor exceeds the flux of water molecules becoming liquid. Condensation occurs if the flux of water molecules becoming liquid exceeds the flux of water molecules becoming vapor. Surface water temperature largely controls the rate of evaporation, because more energetic water molecules (in warmer water) escape a water surface more readily than less energetic water molecules (in colder water).

A similar type of two-way exchange of water molecules takes place at the interface between ice (or snow) and air. Sublimation occurs when more water molecules

INVESTIGATING CLIMATE AND WEATHER

become vapor than solid, and deposition occurs when more water molecules become ice than vapor.

Water vapor enters the atmosphere mostly by evaporation and sublimation of water at the Earth's surface along with transpiration of water by plants. There is an upper limit to the water-vapor component of air. This limit depends largely on temperature. Air is saturated when the water vapor component of air is at its upper limit. The saturation concentration of air increases with temperature, so that warm saturated air has more water vapor than cold saturated air. It follows that sufficient cooling of unsaturated air causes it to become saturated. When air is saturated, excess water vapor condenses (or deposits) into clouds. This happened when you breathed into the cold metal container in the cooler at Station 2.

The relative humidity is defined as the ratio of the actual amount of water vapor to the amount of water vapor at saturation. It is always expressed as a percentage. Suppose that a 1-kg sample of air at 20°C (68°F) has 7.5 g of water vapor. At that temperature a 1-kg sample of air would be saturated if it had about 15 g of water vapor. Hence, the relative humidity of the sample is (7.5 g/15 g) × 100% = 50%. When air is saturated, its relative humidity is 100%. As unsaturated air is cooled, its relative humidity increases. At a relative humidity of 100% water vapor condenses into liquid water droplets or deposits into ice crystals.

Investigation 5: The Causes of Weather

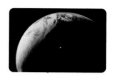

If you fill a glass with ice water on a warm and humid day, small drops of water soon appear on the outside surface of the glass. That water did not leak through the glass. It condensed from the air. The relatively cold surface of the glass chilled the air in contact with the glass causing its relative humidity to increase to 100%. At saturation, water vapor condensed to the small liquid drops. Dew or frost on the grass on a chilly morning forms in the same way, when the ground surface is chilled by radiating its heat out to space on a clear night.

Clouds and Precipitation

When air is heated in a rigid closed container, it tries to expand, because air molecules move faster when the air is warmer. The molecules collide more and more strongly with the walls of the container. This increases the pressure on the inside surfaces of the container. The atmosphere has no walls (except the Earth's surface) and is free to expand when heated. For this reason, at constant pressure (at the Earth's surface, for example) the density of air increases with falling temperature. The density of air decreases with rising temperature. The balloons at Station 3 are flexible, so that the air inside was free to expand and contract in response to changes in temperature.

If air is heated near the ground, it expands. It is then less dense than the air nearby. The cooler, denser air nearby

INVESTIGATING CLIMATE AND WEATHER

pushes under the warmer, less dense air, causing it to rise in the atmosphere. Earlier in this module you learned that the air pressure decreases upward in the atmosphere. As the air rises, it expands in response to the falling pressure. As the air expands, it cools. The reason for the cooling is that the air uses some of its thermal energy doing the work of pushing on the surrounding air. That leaves the air with less thermal energy; in other words, its temperature falls.

Most clouds consist of water droplets that have condensed from water vapor in the air. The droplets fall very slowly toward the Earth. Larger droplets fall faster than smaller droplets. When a larger droplet catches up with a smaller droplet on the way down, the two combine to form an even larger droplet. That one then falls even faster, and sweeps up even more small droplets. Soon a large drop is formed, which falls to Earth' surface as rain.

Raindrops that fall through very cold air near the Earth's surface can freeze to form little grains of ice called ice pellets or sleet. Snowflakes, however, are not frozen raindrops. Snowflakes grow in clouds by addition of water molecules onto their crystal surfaces, directly from the water vapor of the surrounding air—a process called deposition.

Investigation 5: The Causes of Weather

Review and Reflect

Review

1. Describe briefly what you discovered at each station.

Reflect

2. Explain why the global water cycle is called a cycle. You may wish to use a diagram to illustrate your answer.
3. What is the role of heat energy in weather? Give some examples.
4. Make a list of all of the different kinds of pathways you can think of that a water molecule might follow as it goes through the global water cycle. Remember that the water molecule can exist as water vapor, liquid water, or ice, and that it can occur at or near the Earth's surface.

Thinking about the Earth System

5. What role do plants play in the global water cycle?
6. What controls the part of precipitation that runs off into river and stream channels versus the part that infiltrates into the ground?
7. In what sense does the global water cycle involve the flow of energy as well as water?

Thinking about Scientific Inquiry

8. How did you use modeling in this investigation?
9. Why do you think sharing findings is an important process in scientific inquiry?

INVESTIGATING CLIMATE AND WEATHER

Investigation 6:

Climates

Key Question

Before you begin, first think about this key question.

What is the difference between weather and climate?

Think about what you have learned about the nature of weather and how it is reported. How does it differ from climate?

Share your thinking with others in your class. Keep a record of the discussion in your journal.

Materials Needed

For this investigation, all groups will need:

- a blank global map
- colored pencils or markers
- three heat-resistant containers with a pencil-size hole punched in the center of each lid (three per group)
- water supply
- sand
- three thermometers
- heat lamp
- graph paper
- climate resources (books, CD-ROMs, Internet access, etc.)
- poster board and presentation supplies

Investigate

Part A: Climatic Regions of the World

1. Look at the map on the following page showing the various climatic regions of the world. Read over the names of the types of climate.

   a) What weather elements do the name of the climates imply?

Investigation 6: Climates

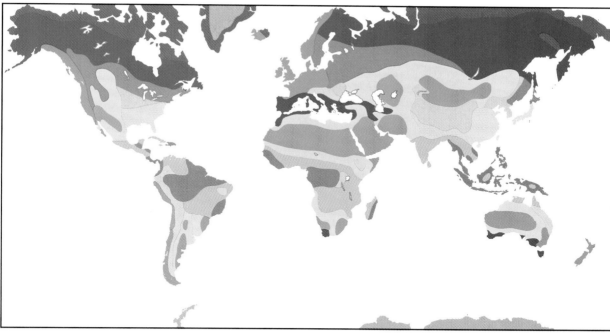

- Tropical Rain forest, Monsoon (wet)
- Tropical Savanna (wet summer, dry winter)
- Steppe (semi arid)
- Desert (dry)
- Mediterranean (dry summer)
- Humid Subtropical
- Marine West Coast (no dry season)
- Humid Continental, Warm Summer
- Humid Continental, Cool Summer
- Subarctic (snow climate)
- Tundra (cold ice climate)
- Ice Cap

b) What does this tell you about how climate is defined?

c) Write down any other questions you may have about the map, its legend, and the climatic regions of the world.

2. Find the area on the map where you live.

With your group, discuss any ideas that you might have about the weather where you live. Write down your ideas.

a) What is it usually like in the summer? (Mostly high temperatures or mostly low? Mostly wet or mostly dry?)

b) What is it usually like in the winter?

c) What kinds of plants and animals live in your area?

d) How do people adapt to changes in the seasons?

3. Make a class profile of your ideas about your local climate. Think carefully and record your answers to the questions on the following page.

Inquiry
Scientific Questions

Scientific inquiry starts with a question. Scientists take what they already know about a topic, then form a question to investigate further. The question and its investigation are designed to expand their understanding of the topic. You are doing the same.

Investigating Earth Systems

INVESTIGATING CLIMATE AND WEATHER

   a) What is it about your area that helps vegetation or crops to grow well, or not well?

   b) What is it about your area that makes it suitable for the animals that live there?

   c) What kind of clothing do people wear in winter in your area? In the summer?

   d) At what latitude do you live?

   e) Are you close to a large body of water (e.g., ocean, Great Lakes), or are you landlocked?

   f) What is your elevation above sea level?

4. Find other parts of the world that have the same climate as your area.

   a) Shade these regions in on a blank copy of the map of the world.

   b) What do all of the areas have in common?

   c) What other factors might affect areas that have your climate?

   d) Would you expect to find the same climate as yours in the polar regions? Why or why not?

5. As you saw from your climate map, the two main weather elements that are used to describe climate are air temperature and precipitation.

   a) Why do you think it is that some areas of the world are typically hot and others are typically cold?

   b) Why are some areas typically dry and others typically moist?

Part B: Ability of Different Materials to Hold Heat

1. One factor related to weather and climate is how different forms of matter that make up the Earth (solids, like soil; liquids, like water; and gases, like air) hold the heat from the Sun.

   Work together in your group to design an experiment to investigate what happens when these three kinds of matter are heated and left to cool.

Be sure that your teacher checks your procedure before you begin.

Investigation 6: Climates

You will be using the following materials:
- heat-resistant containers with lids that have a thermometer-size hole in the center;
- sand, water, and air;
- thermometers;
- heat lamp.

Make sure that the experiment you design is "fair" (objective and systematic).

Complete your design. Decide on the steps you will take from start to finish. Be sure to include any safety precautions you will take.

As a group, decide what you think will happen.

a) Record your group's prediction and the reason for the prediction. This is your hypothesis.

b) Include your procedure in your journal.

c) In your journal create a table for recording your data.

2. With the approval of your teacher conduct your experiment.
   a) Record all your observations.
   b) Graph your data and look for any patterns that emerge.
   c) Which substance had the highest beginning temperature? Why do you think that was?
   d) Which substance cooled most slowly? most quickly? What ideas do you have about that?
   e) How can what you just observed about land, water, and air help you to understand some of the factors that affect climate?

Inquiry

Hypotheses

When you make a prediction and give your reasons for that prediction, you are forming a hypothesis. A hypothesis is a statement of the expected outcome of an experiment or observation, along with an explanation of why this will happen.

A hypothesis is never a guess. It is based on what the scientist already knows about something. A hypothesis is used to design an experiment or observation to find out more about a scientific idea or question. Guesses can be useful in science, but they are not hypotheses.

Dependent and Independent Variables

In all experiments, there are things that can change (vary). These are called variables. In a "fair" test, scientists must decide which things will be varied in the experiment and which things must remain the same. In this investigation you will make measurements to determine how well each kind of matter holds heat. This is the dependent variable. The kind of matter that you are testing is called the independent variable. All other variables must be controlled; that is, nothing else should change.

INVESTIGATING CLIMATE AND WEATHER

Part C: Creating Climate Clues

1. To help you learn about climates, each group in your class will specialize in researching a particular climate type.

   Choose a particular climatic zone on the map.

   You are going to prepare a set of six clues that your classmates will need to use to guess your climate type.

   Clues that you can give about your climate can include:
   - graphs of monthly temperature and precipitation for the climate;
   - descriptions of what people are wearing on a typical day;
   - nearby bodies of water;
   - well-known landforms;
   - typical animals that live there;
   - crops that are grown in the area.

2. Use all the resources you have available to conduct your research and make up your clues.

3. When all the clues are ready, each group in the class gets the chance to present its clues to the other groups.

4. To participate, someone from another group must raise his or her hand. That person then gets a chance to guess what the climate is. The person trying to guess must also give a good reason why he or she thinks your clue fits with a certain climate.

5. Continue until all groups have a chance to present their climate clues.

   a) Which clues were the most useful in guessing the climates, and why?

   b) Which resources were most useful, and why?

Investigation 6: Climates

Digging Deeper

WEATHER AND CLIMATE
The Difference between Weather and Climate

Weather is the state of the atmosphere at a particular place and time described in terms of temperature, air pressure, clouds, wind, and precipitation. Climate is the long-term average of weather. It is observed over periods of many years, decades, and centuries. In many areas of the United States, the daily high temperature or the daily low temperature can vary by as much as 30°F from one day to the next. In contrast, the average temperature for a whole year seldom varies by more than 1°F.

Factors That Determine the Climate

The two most important factors in describing the climate of an area are temperature and precipitation. The yearly average temperature of the area is obviously important, but the yearly range in temperature is also important. Some areas have a much larger range between highest and lowest temperature than other areas. Likewise, average precipitation is important, but the yearly variation in rainfall is also important. Some areas have about the same rainfall throughout the year. Other areas have very little rainfall for part of the year (the dry season) and a lot of rainfall for the other part of the year (the wet season).

As You Read…
Think about:
1. What is the difference between weather and climate?
2. What factors determine the climate?
3. How does average yearly temperature vary with latitude?
4. How does precipitation affect vegetation?
5. Why can local climate vary over very short distances?

INVESTIGATING CLIMATE AND WEATHER

The average temperature in an area depends mainly on the latitude. Generally, areas near the Equator have high average temperatures, and areas nearer the North and South Poles have lower average temperatures. The range of temperature, however, depends more on where the area is located in relation to the ocean. Areas where winds usually blow from the ocean have a smaller range of temperature than areas far away from the ocean, in the interior of a continent. That is because water has a much greater heat capacity than rock and soil, as you saw in your investigation. It takes much more heat from the Sun to warm up water than it takes to warm up rock and soil. Likewise, water cools off much more slowly than rock and soil on cold, clear nights and in the winter.

Climate and Vegetation

The plant community in an area is the most sensitive indicator of climate. Areas with moderate to high temperatures and abundant rainfall throughout the year are heavily forested (unless humans have cleared the land for agriculture). Areas with somewhat less rainfall are mainly grasslands, which are called prairies in North America. Humans have converted grasslands into rich agricultural areas around the world. Even in areas with high yearly rainfall, trees are scarce if there is not much rainfall during the warm growing season. As you know, regions with not much rainfall and scarce vegetation are called deserts, or arid regions. Areas with somewhat greater rainfall are called semiarid regions. The major problem with using semiarid or arid regions for agriculture is that ground water is removed to irrigate crops. In many cases, the removal of

Investigation 6: Climates

ground water exceeds the rate at which it is naturally replaced (by precipitation that reenters the ground water system). As a result, water resources are lost, water levels in wells fall, and less ground water flows to recharge streams and rivers, which has other impacts on the Earth system. Another problem with semiarid regions is that when humans use them for agriculture, the loss of natural vegetation can cause the areas to become deserts.

Microclimate

It is easy to understand how climate can vary over very large areas, because of slight changes in temperature or rainfall. Climates can also vary over very short distances. Local differences in climate are described by the term "microclimate." Differences in microclimate might explain some of the differences in weather from place to place you likely observed in the first investigation in this module.

Sometimes, low-lying areas are colder at night than higher ground nearby. On clear nights, the ground is chilled as its heat is radiated out to space. The cold ground then chills the air near the ground. The chilled air is slightly denser than the overlying air, so it tends to flow slowly downhill, in the same way that water flows downhill. The cold air "ponds" in low areas. These are places where the first frosts of autumn are earliest and where the last frosts of spring are latest. If you ever have a chance to plant fruit trees, plant them on the highest ground around!

In hilly areas, north-facing slopes get less sunshine than south-facing slopes. Local temperatures on the north-facing slopes are colder than on south-facing slopes in both summer and winter. In areas with winter snows, the snow melts much later on north-facing slopes.

INVESTIGATING CLIMATE AND WEATHER

Review and Reflect

Review

1. Which climate type has the most rainfall? Which has the least?
2. Which climate type has the shortest growing season? Which has the longest?
3. How and why do oceans and continents affect climate?

Reflect

4. Does climate influence human population size in an area? If so, how?
5. Imagine that the climate in your area changed suddenly.
   a) What would have to be done to homes and other buildings?
   b) How would this affect what you wear?
   c) How would this affect what you eat?
   d) How would it affect what you could grow in a garden?
   e) How would it affect transportation?
   f) How would it affect the work people do?
   g) How would it affect what you do for recreation?

Thinking about the Earth System

6. On your *Earth System Connection* sheet, note how the things you learned in this investigation connect to the geosphere, hydrosphere, atmosphere and biosphere.
7. Global warming would cause sea level to rise. How would this affect other Earth systems? What would be the effect? Higher sea level means a higher base level for rivers. How might this affect erosion by rivers?

Thinking about Scientific Inquiry

8. How did you use questions to answer by inquiry in this investigation?
9. How is a hypothesis different from a guess?
10. What factors must be considered when designing a "fair" experiment?

Investigation 7: Exploring Climate Change

Investigation 7:
Exploring Climate Change

Key Question
Before you begin, first think about this key question.

What evidence suggests that climate has changed in the past?

Think about what you have learned so far. Is global climate changing? How do you think scientists learn about what the global climate was like before weather data were recorded? What "climate clues" are out there?

Share your thinking with others in your class. Keep a record of the discussion in your journal.

Materials Needed

For this investigation your group will need:

- resources on global-climate change (books, CD-ROMs, Internet access, etc.)

Investigate

1. Studying weather data is only one way of learning about climate and how it changes. Paleoclimatologists are climatologists who study evidence from the past (ice cores, ocean bottom cores, tree rings, rocks, and fossils, among others) to find out more about climate in the past.

INVESTIGATING CLIMATE AND WEATHER

Look at the climate evidence that follows.

Also look at the climate map of the world in the previous investigation.

Read the information about the items in the pictures (fossils, tree rings, etc.) to find out more about climate in the past.

Discuss the questions with your group.

Keep a record of your group's discussion in your journal.

Climate Evidence 1

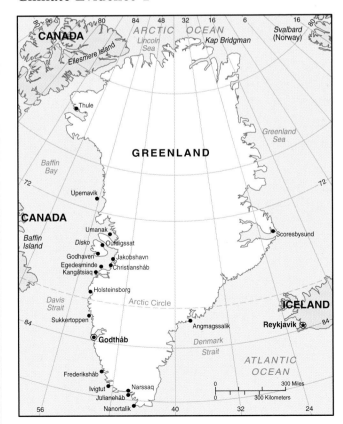

Investigation 7: Exploring Climate Change

Fossil shells, found on the coast of Greenland, show that certain species of warm-water mollusks lived there about 8500 years ago.

a) What is the climate like in Greenland today?

b) What clues do you think these fossils give about climate changes in Greenland?

c) What ideas do you have about what might have caused climate changes in Greenland?

Climate Evidence 2

Inquiry

Considering Evidence

Scientific judgments depend on solid, verifiable evidence. Claims by scientists that the global climate is changing must be supported by evidence. You must be certain about the reliability of any evidence that you use. You must also know how the accuracy of the evidence can be checked.

A fossilized impression of a banana, 43 million years old, was found in the state of Oregon. Find some of the places on the world climate map where bananas grow in the world today.

a) What is the climate like in Oregon today?

b) What clues can the fossil give you about climate changes in Oregon?

c) What ideas do you have about what might have caused climate changes in Oregon?

INVESTIGATING CLIMATE AND WEATHER

Climate Evidence 3

Fossils found near Antarctica include a meat-eating dinosaur that lived 200 million years ago. The dinosaur preyed on small animals that, in turn, fed on lush plants.

a) What is the climate like in Antarctica today?

b) What did the climate need to be like for the dinosaur to survive?

c) What clues can the fossils give about climatic changes in the area around Antarctica?

Climate Evidence 4

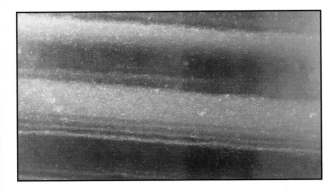

Ice cores can be taken wherever very deep ice exists. Some of these ice cores have sampled ice that is more than 400,000 years old. Ice traps air and dust in tiny bubbles in the ice. Climatologists can study these materials to find out about past climates.

a) How many darker layers and how many lighter layers can you see in this example?

b) What do the layers suggest about the climate during the time period that was sampled?

Investigation 7: Exploring Climate Change

Climate Evidence 5

Cave paintings like this have been found in the Sahara Desert. They have been dated as far back as 4000 B.C. Observe the kinds of animal it shows.

a) What is the climate like in the Sahara today?

b) What did the climate need to be like for the animal in the painting to live in the Sahara?

Climate Evidence 6

This picture shows the tree growth rings from a bristlecone pine tree. Some of these trees live to be as old as 4000 years. A wide ring means warm, humid weather; a narrow ring means cold, dry weather.

a) Count the number of growth rings on this example (approximately the number of years the tree has lived).

b) What evidence do you have about past climate, on the basis of the tree rings pictured?

INVESTIGATING CLIMATE AND WEATHER

2. When you have finished working with all of the evidence, look over what you have written about climate in the past. Consult any resources you have available about climate in the past to add more information to what you already know.

   a) Write down any new ideas you discover.

3. As a class, pull all your investigations together. Discuss your answers and come to a consensus as a class.

   a) List all the evidence you have been able to find that shows how climates have changed over time. Wherever possible, include the time scales involved.

   b) What evidence do you now have that shows that climate change happens slowly? What evidence do you now have that climate change can happen rapidly?

As You Read...
Think about:
1. What is a climate proxy?
2. How are ice cores used to determine past climates?
3. In your own words, explain the astronomical theory of the ice age.
4. What are two events that can affect worldwide climate over a period of a few years?

Digging Deeper

MEASURING CLIMATE CHANGE
Climate Proxies

The Earth's climate has changed greatly through the billions of years that constitute geologic time, and even in recent centuries. The study of past climates is called paleoclimatology (*paleo-* means early or past).

Something that represents something else indirectly is called a proxy. There are many proxies for past climate. They provide a lot of information, although none is perfect. Some, like kinds of past plants and animals, are easy to understand. Some important proxies, involving the chemical element oxygen, are more difficult to understand.

Ice Cores

Thin cores of ice, thousands of meters deep, have been drilled in the ice sheets of Greenland and Antarctica. They are preserved in special cold-storage rooms for study.

Investigation 7: Exploring Climate Change

Glacier ice is formed as each year's snow is compacted under the weight of the snows of later years. A slightly darker layer that contains dust blown onto the ice sheet during summer, when not much new snow falls, marks each year's new ice. The winter layer consists of cleaner and lighter-colored ice. The layers are only millimeters to centimeters thick. They can be dated by counting the yearly layers. The oxygen in the water molecules also holds a key to past climate. Scientists are able to use the oxygen atoms in the glacier ice as a proxy for air temperature above the glacier.

Past Glaciations

Ice sheets on the continents have grown and then shrunk again at least a dozen times over the past 1.7 million years. Many climate proxies make that very clear. Deposits of sediment and distinctive landforms left by these glaciers are present over large areas of North America and Eurasia. Proxies for global temperature show gradual cooling as the ice sheets form. They also then show very rapid warming as the ice sheets melt back. Intervals of relatively high temperature between glaciations are called interglacials. Past interglacials have lasted about 10,000 years. Civilization developed only within the last interglacial—and you are still in it! The graph shows the estimated global surface temperature for the last 420,000 years.

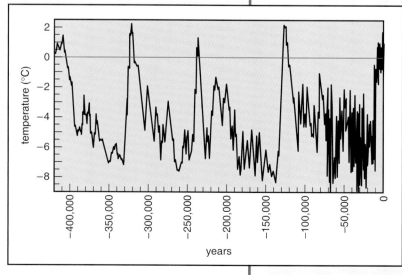

INVESTIGATING CLIMATE AND WEATHER

This was obtained from the longest and most informative ice core. This core was taken at the Vostok station in Antarctica by a team of Russian, American, and French scientists.

Causes of Climate Change

Probably the two most important factors that determine Earth's climate are the amount of heat the Sun delivers to the planet, and also where the continents are located relative to the Equator. Continental ice sheets cannot develop unless plate tectonics cause one or more continents to be at high latitudes.

The Earth revolves around the Sun once a year. Its orbit is in the shape of an ellipse. If the Earth were the only planet, its orbit around the Sun would be almost unchanging. The pull of the other planets on the Earth causes the Earth's orbit to be much more complicated. The orbit changes slightly in several different ways. These changes occur over periods that range from about 20,000 years to about 100,000 years. They cause slight differences in how much of the Sun's heat the Earth receives in winter versus summer and at high latitudes versus low latitudes.

One theory holds that the small changes in the Earth's orbit trigger the advance and retreat of ice sheets. This

Investigation 7: Exploring Climate Change

theory is known as the astronomical theory of the ice ages. It was first developed in the 1920s and 1930s by the Serbian astrophysicist Milutin Milankovitch (1879–1958). It was not widely accepted by the scientific community until the 1970s. Although most scientists today accept this theory, details of how the changes govern the volume of ice sheets are still only partly understood. For example, the extremely fast melting of the ice sheets, compared to the long times needed for them to form, is still a mystery.

The astronomical theory is only part of the story. Climate is known to change on time scales as short as a century or even a few years, and the cause (or causes) of these changes are still not clear.

Some violent volcanic eruptions are known to influence climate on a scale of one to two years. Substances from the volcano are blasted into the air. Some are so small that they can remain in the atmosphere for many months to a few years. While there they both absorb and reflect solar radiation. This reduces the amount of sunlight that reaches Earth's surface. A cooling of up to 1°C at Earth's surface may be observed.

El Niño and La Niña are large-scale air—sea interactions in the tropical Pacific. They can also affect the climate in many regions of the world over periods of one to two years. Changes occur in the sea-surface temperature. This affects the air pressure patterns in the tropical Pacific. As a result, storm tracks at middle and high latitudes are altered. Some regions that are usually wet have droughts. Other regions that are arid or semiarid have heavy rains.

INVESTIGATING CLIMATE AND WEATHER

Review and Reflect

Review

1. Describe at least three ways that scientists can detect or measure climate change.
2. What are some of the possible causes of climate change?

Reflect

3. What kinds of evidence suggest that climates have changed over time?
4. How convincing does this evidence seem to be?
5. What further evidence is needed to answer some of the questions about climate change?

Thinking about the Earth System

6. On your *Earth System Connection* sheet, note how the things you learned in this investigation connect to the geosphere, hydrosphere, atmosphere, and biosphere.

Thinking about Scientific Inquiry

7. What are some sources of error associated with the ways of detecting climate change that are described in this investigation?

Investigation 8: Climate Change Today

Investigation 8:
Climate Change Today

Key Question

Before you begin, first think about this key question.

How is the global climate changing?

Think about what you have learned about climate change. Do you think that the world's climate is changing? If so, what are the prospects for the future? What will the climate be like for you, your children, and your grandchildren?

Share your ideas with your classmates. Record your thoughts in your journal.

Materials Needed

For this investigation, your group will need:

- your weather data
- weather data for your area for the past 30 years
- resources on global-climate change (books, CD-ROMs, Internet access, etc.)
- presentation materials (poster board, markers, etc.)

Investigate

1. Pull together all the weather data and climate information that you have collected over the course of the module.

INVESTIGATING CLIMATE AND WEATHER

Examine the climate of your area and compare it to the weather data you have collected.

Also compare the climate with the weather data from your area for the past 30 years.

Discuss what patterns and relationships you notice in your climate and weather data with your group.

a) What trends, if any, do you notice in the weather data over the past 30 years?

b) Do the temperatures appear to be increasing, decreasing, or staying about the same? How can you tell?

c) What is the average yearly rainfall in your area?

d) On the average, have you had more or less rainfall than one would predict for your climate over the past ten years?

e) How could you explain what you observe?

2. Now think about what the future climate might be like in the region where you live.

Think about these questions:

- Will it be much warmer or much cooler than it is now? If so, how much warmer or cooler? Or will the climate be about the same as now?

- Do you think there will be much more rainfall per year, much less, or about the same? Why do you think that?

a) Write down a prediction for what you think the climate will be like where you live 100 years from now. Also include the reasons for your prediction.

3. Discuss your predictions with other people in your class.

a) List the predictions and reasons that others have.

4. For this investigation you may decide to stay in your normal groups or to work with people who have ideas similar to yours. Whichever you decide, your next task is to look at the individual predictions and the reasons you gave for them.

In your group, develop a final prediction, with reasons, that you all agree on.

a) Write this down clearly. It will be the key starting point for your research.

Investigation 8: Climate Change Today

5. Begin by assessing what information you already have available, then discuss what further information sources you need to consult.

   Keep in mind that you need to look for data that both do and do not support your predictions, and the reasons you gave for them.

   In your search for new information, you might want to divide the tasks. For example, one person could be responsible for exploring the Internet, another for locating and investigating text books, another looking at trade books, and so on. Also, do not forget to consult each group member's journal for information.

   When you have a plan, make a schedule for the information gathering part of your research project.

   a) Record your schedule in your journal.

6. Once you have gathered all the information you can, look at what you have. In your group discuss these questions:

   - What evidence is there from all of this information that is relevant to the research prediction, and the reasons you gave for it?
   - Which parts of the evidence support the prediction, which do not support it, and which do neither but are still important?

   It may be helpful to construct a chart within which to put these various pieces of evidence. You may find an alternative way to do this. By now you have enough experience of dealing with data to decide.

   As well as using this way of sorting your information, keep in mind that you might later want to include data charts, or other forms of representation, in your final report. If one person in your group has special talents in designing these things, give this job to him or her.

7. When you have fully reviewed all your collected evidence and information, you need to analyze it against your prediction.

   This is a very important step. Here you need to decide whether your prediction is supported by the evidence, not supported, or supported somewhat but not enough to be conclusive.

Investigating Earth Systems

INVESTIGATING CLIMATE AND WEATHER

Together, work on this analysis and then reach agreement on how both your prediction and the reasoning behind it are reflected by the evidence. When this is complete, and all are agreed, your research phase is complete.

8. Decide on a way of presenting the results of your research for others to see and understand. This is going to be your research report. This must include:

   - your predictions and the reasons for them;
   - your evidence for or against your predictions;
   - your conclusions about whether the global climate is changing or not, based upon the evidence you have found and analyzed;
   - what inquiry processes you used in your research and how you used them;
   - how your research relates to the Earth system.

   Together, discuss how you will do this in your report. Decide what sections your report will contain, where illustrations and data representations will be included, and how you will present the written information.

9. Create your research report and share it with the class. Be sure to study carefully each other group's research reports.

   a) What did you learn from the other reports?

   b) How much agreement is there between the various reports on global change? Explain.

Investigation 8: Climate Change Today

Digging Deeper

GLOBAL CLIMATE CHANGE

The Earth seems to be getting warmer. Graph 1 shows how yearly average temperature has changed since 1880. This is about the time when temperatures first began to be recorded in an organized way at weather stations around the world. The curve on the graph has lots of ups and downs, but there is a clear upward trend. Graphs like this have some uncertainties, however. For example, there is a problem about how to adjust the curve to take into account the growth of large cities, where so many of the weather stations are located. Urbanization causes temperature readings in large cities to be somewhat higher than in surrounding rural areas. The upward trend shown is almost certainly real, though. Other evidence, like the general shrinking of glaciers all around the world and the thinning of the Arctic ice pack, tell the same story.

**As You Read...
Think about:**

1. How has yearly average temperature changed since 1880?
2. How has global average temperature changed during the past 2400 years?
3. What is Earth's principal greenhouse gas?
4. Why has carbon dioxide increased over the past decades?
5. Why are computer models limited in their predictions of Earth's future climate?

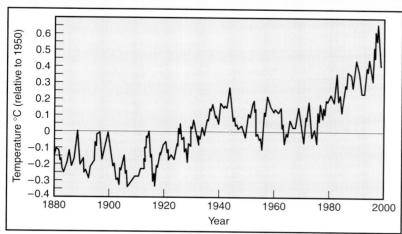

Graph 1

Graph 2 is similar to Graph 1. It shows global average temperature for the past 2400 years. The curve is less certain than the one in Graph 1, because it is based on various proxies for temperature. You learned

INVESTIGATING CLIMATE AND WEATHER

about these in the previous investigation. The highest temperatures are about 1°C above 20th century mean temperatures. The lowest temperatures are about 1°C below 20th century mean temperatures.

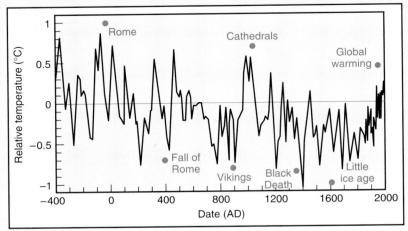

Graph 2

Here's a big question, and an important one. Has the increase in temperature since the beginning of the 20th century been caused by human activity, or is it just another natural upward "spike" like several during the past two millennia, shown in Graph 2? Most climatologists think that the upward trend in temperature during the 20th century is at least partly caused by human activity.

Several gases in the Earth's atmosphere are called greenhouse gases. The most important is water vapor, but carbon dioxide is important also. Early on in Earth's history, carbon dioxide was the principal greenhouse gas. Carbon dioxide has always been part of the Earth's atmosphere, but has been increasing more and more rapidly in recent times (Graph 3). Coal, oil, and natural gas are called fossil fuels, because they come from plant and animal material that was buried in the Earth's sediments. When they are burned, they add carbon dioxide gas to the atmosphere. Along with the other greenhouse gases, carbon dioxide absorbs some of the heat that the Earth's surface sends

Investigation 8: Climate Change Today

out to space and radiates heat back to the surface. That increases the Earth's average surface temperature. The effect is the same as with the glass of a greenhouse, although the process itself is not exactly the same.

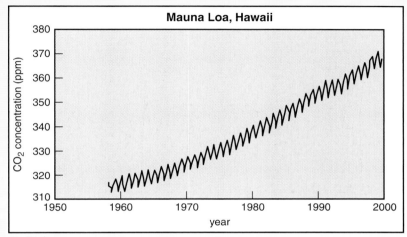

Graph 3

Several groups of climatologists have been developing computer models of the Earth's climate. They try to build in all of the important controls of climate into the model. Then they start the model with the present climate and let it run to see what the future climate will be. The models are not perfect, because it is very difficult to simulate some of the most important influences on climate. The behavior of clouds is especially tricky to model. The models have one thing in common, though. They predict that the Earth's surface temperature is likely to increase by as much as 2°C between 2000 and 2050 if the upward trend in

INVESTIGATING CLIMATE AND WEATHER

atmospheric carbon dioxide continues. A look back at Graph 2 shows that a rise of 2°C would make the Earth's temperature much higher than during even the warmest periods in human history.

If the predictions about global warming come true, many things about the Earth's climate, aside from just temperature, are likely to change. Some regions will get more rainfall, and other regions will get less. The frequency and intensity of severe storms are likely to increase. As the world's glaciers continue to melt, sea level around the world will rise, by as much as half a meter or so. That might not sound like a lot to you, but think of the flooding that it would cause in coastal cities and islands around the world!

Will the predictions of the computer models come true? That is likely, although not certain. All that science can do is try to make likely predictions. How to act upon the predictions is for human society to decide.

Review and Reflect

Reflect

1. How did your ideas about how global climate will change in the future differ from those of other groups? Explain why there might be differences of opinion.

Thinking about the Earth System

2. Add any new connections that you have found between climate change and the Earth system (biosphere, atmosphere, hydrosphere, and geosphere) to your *Earth System Connection* sheet.

Thinking about Scientific Inquiry

3. What evidence did you find for future climate change?